The Science of Computer Benchmarking

SOFTWARE · ENVIRONMENTS · TOOLS

The series includes handbooks and software guides, as well as monographs on practical implementation of computational methods, environments, and tools. The focus is on making recent developments available in a practical format to researchers and other users of these methods and tools.

Software, Environments, and Tools

Michael W. Berry and Murray Browne, *Understanding Search Engines*
Jack J. Dongarra, Iain S. Duff, Danny C. Sorensen, and Henk A. van der Vorst, *Numerical Linear Algebra for High-Performance Computers*
R. B. Lehoucq, D. C. Sorensen, and C. Yang, *ARPACK Users' Guide: Solution of Large-Scale Eigenvalue Problems with Implicitly Restarted Arnoldi Methods*
Randolph E. Bank, *PLTMG: A Software Package for Solving Elliptic Partial Differential Equations, Users' Guide 8.0*
L. S. Blackford, J. Choi, A. Cleary, E. D'Azevedo, J. Demmel, I. Dhillon, J. Dongarra, S. Hammarling, G. Henry, A. Petitet, K. Stanley, D. Walker, and R. C. Whaley, *ScaLAPACK Users' Guide*
Greg Astfalk, editor, *Applications on Advanced Architecture Computers*
Roger W. Hockney, *The Science of Computer Benchmarking*
Françoise Chaitin-Chatelin and Valérie Frayssé, *Lectures on Finite Precision Computations*
Richard Barrett, Michael Berry, Tony F. Chan, James Demmel, June Donato, Jack Dongarra, Victor Eijkhout, Roldan Pozo, Charles Romine, and Henk van der Vorst, *Templates for the Solution of Linear Systems: Building Blocks for Iterative Methods*
E. Anderson, Z. Bai, C. Bischof, J. Demmel, J. Dongarra, J. Du Croz, A. Greenbaum, S. Hammarling, A. McKenney, S. Ostrouchov, and D. Sorensen, *LAPACK Users' Guide, Second Edition*
Jack J. Dongarra, Iain S. Duff, Danny C. Sorensen, and Henk van der Vorst, *Solving Linear Systems on Vector and Shared Memory Computers*
J. J. Dongarra, J. R. Bunch, C. B. Moler, and G. W. Stewart, *LINPACK Users' Guide*

The Science of Computer Benchmarking

Roger W. Hockney

siam

Society for Industrial and Applied Mathematics
Philadelphia

10 9 8 7 6 5 4 3 2

Library of Congress Cataloging-in-Publication Data

Hockney, Roger W., 1949–1999
The science of computer benchmarking / Roger W. Hockney.
p. cm. - - (Software, environments, tools)
Includes bibliographical references and index.
ISBN 0-89871-363-3 (pbk.)
1. Electronic digital computers - - Evaluation. I. Title II. Series.
QA76.9.E94H63 1996
004.2' 4–dc20 95-46486

The author wishes to thank the following for permission to include material from their publications for which they hold the copyright:

John Wiley & Sons, Inc., New York, USA, for material from *Scientific Programming* [39] in Chapters 1, 2, 3 (including Table 3.1), from *Concurrency: Practice and Experience* [51] in Chapter 4 (including Figs. 4.2, 4.3, 4.4) and from *Scientific Programming* [17, 64] in Chapter 5 (including all Figs. except 5.8).

Elsevier Science Publ. BV (North Holland), Amsterdam, The Netherlands, for material from *Parallel Computing* [46] in Chapter 1 (including Tables 1.1, 1.2, and Fig. 1.1), Chapter 2 (part section 2.6), Chapter 3 (including Fig. 3.3) and Chapter 4 (part section 4.1).

ASFRA, Edam, The Netherlands, for material from *Supercomputer* [47] in Chapter 2 (including Figs. 2.1 to 2.5).

Cover photograph by Jill Davis.

To Mummy, Cedra and Judith

Contents

Foreword

Underlying Roger Hockney's selection of the title for this book is the tacit acknowledgment that much of what has been done in the the world of supercomputer benchmarking to date is not very scientific. It is indeed curious that scientists who presumably have been drilled in the methodology of rigorous scientific inquiry for their particular discipline, whether it be computer science, physics, chemistry or aeronautical engineering, often are very casual about this methodology when reporting on the performance of the computers they employ in their research.

After all, one would expect that any field styling itself as a "science" would employ, and to some extent enforce, the accepted principles of rigorous inquiry:

1. Insuring objectivity in experimental measurements and written reports.
2. Employing effective controls in experiments to isolate what one wants to measure.
3. Carefully documenting environmental factors that may affect experimental results.
4. Providing enough detail in written reports to permit other researchers to reconstruct one's results.
5. Employing standard, unambiguous notation.
6. Comparing one's own results with other results in the literature.
7. Developing mathematical models that accurately model the behavior being studied.
8. Validating these models with additional experiments and studies.
9. Reasoning from these models to explore fundamental underlying principles.

Those who have worked in this field for the past few years do not need to be reminded that these principles have not always been followed, either by computer vendors or application scientists. Indeed, the high performance computing field seems to have acquired a somewhat tawdry reputation, particularly for its frequent use of exaggeration and hyperbole.

This book is a much needed step in the direction of establishing supercomputer benchmarking as a rigorous scientific discipline. Professor Hockney carefully defines underlying terms, describes tools that can measure fundamental underlying parameters, develops mathematical models of performance, and then explores the implications of these models. There is much to learn here, both for those who specialize in this area as well as the general scientist who relies on computation in his/her work.

Hopefully this book is only a start, and others will pick up the reins where Roger has left off. Then perhaps one day we will not have to be reminded that computer performance analysis is a science.

David H. Bailey
NASA, Ames Research Center
Moffett Field, California, September 1995

Preface

The idea for this book arose out of the need for teaching material to support tutorials on computer benchmarking given by the author at Supercomputing94 in Washington DC (in conjunction with David Bailey of NASA Ames) and HPCN Europe 1995 in Milan (in conjunction with Aad van der Steen from the University of Utrecht). These tutorials coincided with the early years of the Parkbench Committee, an organisation of parallel computer users and vendors founded by Tony Hey (Southampton University) and Jack Dongarra (University of Tennessee) and chaired initially by the author. The objective of this group is to create a set of computer benchmarks for assessing the performance of the new generation of parallel computers, particularly distributed-memory designs which are scalable to a large number of processors and are generally known as MPPs (Massively Parallel Processors) or SPPs (Scalable Parallel Processors), and for which there was a notable lack of accepted benchmarks. This committee's first report, assembled and edited by the author and Michael Berry, was published in 1994 and is generally known as "The Parkbench Report". This report contains a lot of condensed material concerned with the definition of performance parameters and metrics, which could not be fully described in the report. This small book can be considered as a tutorial exposition of the more theoretical aspects of the Parkbench Report, and of other related work.

It was John Riganati (David Sarnoff Research Center, and Tutorials Chairman of Supercomputing94) who suggested I presented a tutorial under the title "The Science of Computer Benchmarking". At first this title seemed too presumptuous, and like other recent names containing unproven assertions, e.g. "High (?) Performance Fortran", was likely to attract ridicule rather than respect, and to hold too many hostages to fortune. However, on reflection, I adopted John's suggestion both for the tutorial and this book, because it attracts attention, and highlights the need for a more scientific approach to the subject. For example, benchmarkers do not even have an agreed set of scientific units and metrics in which to express their measurements. The Parkbench Report and this book suggest an extension to the SI system of physical and engineering units for this purpose. But it is in no way claimed that these suggestions are complete or the best, only that they are a first step in the right direction. I have also included a chapter on the concept of "Computational Similarity" which is not part of the Parkbench Report, but which uses dimensional analysis to interpret and understand the scaling of parallel computer performance in a very general way.

The work reported in this book has been carried out in association with Reading, Southampton and Warwick Universities, and I would like to thank my colleagues at these institutions for their assistance and helpful criticism, particularly Chris Jesshope, Tony Hey, Graham Nudd, Mark Baker, Ivan Wolton, Vladimir Getov, John Merlin, Alistair Dunlop, Oscar Naim, Mark Papiani, Andrew Stratton and Jamal Zemerly. I am also indebted to the many members of the Parkbench committee who have taken part in e-mail debates on benchmarking during the writing and assembly of the Parkbench Report, notably David Bailey (NASA), Charles Grassl (Cray Research), Dave Schneider and Xian-He Sun. Especially helpful discussions on benchmarking that I remember well, have been with Vladimir Getov (Westminster University), Robert Numrich (Cray Research), David Snelling (Manchester University), Aad van der Steen (University of Utrecht), Karl Solchenbach (Pallas), and David Keyes (Old Dominion University). Particular thanks are due to Jack Dongarra, David Bailey, Aad van der Steen, Michael Berry, Brian LaRose and Mark Papiani for permitting me to use extracts from their own work. I am also greatly indebted to James Eastwood and AEA Technology for their support in this project, and for their permission to include performance results from the LPM series of benchmark programs. Similarly thanks are due to John Wiley & Sons, Elsevier Science (North-Holland) and ASFRA Publishers for permission to re-use material from their journals.

Work on the theory of benchmarking cannot be successfully performed in isolation, and has been guided throughout by reference to experimental timing measurements made on many different computer installations. Each installation has its own peculiarities, and the measurements could not have been made without the very willing help of staff – often beyond the call of normal duty – at these installations who are too numerous to name individually, but I would like to thank all those who have helped at Southampton University (SN1000 Supernode, Meiko CS-2) and Parallel Application Centre (iPSC/860), Parsys Ltd (SN1000 Supernode), Sandia National Laboratory (Intel Paragon) and the Maui High-Performance Computer Center (IBM SP2). Performance data from Sandia and Maui were obtained courtesy of the United States Air Force, Phillips Laboratory, Albuquerque, New Mexico.

The whole of this book has been printed from electronic files prepared using the LaTeX text processing system, and I would like to express my thanks to Leslie Lamport for making this excellent software available to the scientific community. I have also used the IBM PC version, EmTex, written by Eberhard Mattes which made it possible to prepare text at home. Thanks Eberhard, and all others associated with EmTex. Many of the figures were prepared using the Sigma-Plot4 graphics package from Jandel Scientific which proved both reliable and easy to use. The short time from book concept in February 1995, through submission of electronic manuscript at the beginning of October, to the book's launch in December at Supercomputing95, San Diego, must be something of a record, and could only have been achieved with the willing help of many at SIAM publishing, Philadelphia, particularly Susan Ciambrano, Bernadetta DiLisi, Corey Gray, Vickie Kearn and Colleen Robishaw. Almost all the communication between author and publisher was conducted rapidly by e-mail over Internet, so that all associated with this project deserve thanks for completely eliminating the problems of postal delays and thereby revolutionising the author/publisher relationship.

Roger Hockney
Compton, Newbury, England, September 1995

Chapter 1

Introduction

1.1 The PARKBENCH Committee

In November 1992, at Supercomputing92, a group of parallel computer users and manufacturers held a "birds-of-feather" session to setup a committee on parallel-computer benchmarking. Under the initiative of Tony Hey and Jack Dongarra and chaired by the author, it was agreed that the objectives of the group were:

1. To establish a comprehensive set of parallel benchmarks that is accepted by both users and vendors of parallel systems.

2. To provide a focus for parallel benchmarking activities and avoid unnecessary duplication of effort and proliferation of benchmarks.

3. To set standards for benchmarking methodology and result-reporting together with a control database/repository for both the benchmarks and the results.

4. To make the benchmarks and results freely available in the public domain

The group subsequently adopted the name PARKBENCH committee standing for PARallel Kernels and BENCHmarks. In the first year it produced a draft report which was distributed at Supercomputing93 and a final version which was published in Parallel Programming in the Summer 1994 edition [39]. This report is referred to here simply as "The Parkbench Report", and it is the purpose of this book to give a tutorial exposition of the main theoretical ideas on performance characterisation and benchmarking methodology from this report. We also discuss briefly below other important benchmarking initiatives, and in Chapter-4 give a theory of performance scaling based on the use of dimensional analysis and the idea of computational similarity which is analogous to dynamical similarity in fluid dynamics.

The Parkbench report is available by anonymous ftp and on two World-Wide Web servers at the Universities of Southampton and Tennessee [15, 76, 57]. It is recommended that the reader obtain a copy from one of these sources to refer to whilst reading this book. Anyone interested in taking part in the activities of the committee, including e-mail discussion of the developing benchmarks and methodology, may request to be included in the e-mail reflector list by sending e-mail to this effect to:

dongarra@cs.utk.edu

Discussion takes place by sending e-mail to:

parkbench-comm@cs.utk.edu

which is then automatically broadcast to all members on the reflector list. The correspondence is also archived and may be retrieved by anyone interested from the Web or by ftp from

netlib2.cs.utk.edu

The above objectives are pretty uncontroversial, but perhaps differ from some other computer benchmarking activities in emphasising the setting of standards in methodology and result reporting, and in insisting that all benchmarks and results be in the public domain. This was because it was clear to the group that there was currently bad and sloppy practice in the reporting of results due to confusion over the definition of certain performance metrics, particularly Megaflops and Speedup. It was also noted that the obvious success of the Dongarra's Linpack benchmark [24] was largely due to the public availability of the benchmark and the results. These issues are both addressed in the report and expanded on in this book.

1.2 The Parkbench Report

The Parkbench Report was the product of the whole committee, assembled and edited by Roger Hockney (chairman) and Michael Berry (secretary). It is laid out in five chapters each of which was the responsibility of an individual member:

1. Methodology (David Bailey, NASA Ames)
2. Low-level Benchmarks (Roger Hockney, Southampton)
3. Kernel Benchmarks (Tony Hey, Southampton)
4. Compact Applications (David Walker, ORNL)
5. Compiler Benchmarks (Tom Haupt, Syracuse)

Methodology covers the definition of a set of units for expressing benchmark results, and standard symbols for them. These are given as extensions to the set of SI [65] units used universally now by scientists and engineers. In particular it recommends that the commonly used symbol MFLOPs be replaced by Mflop, and MFLOPS be replaced by Mflop/s, in order to distinguish clearly between computer flop-count and computer processing rate. It also covers the precise definition of a set of performance metrics and emphatically bans the use of Speedup as a figure-of-merit for comparing the performance of different computers – an all too common and invalid practice. Some of these matters are controversial and are therefore discussed more fully in Chapter-2 of this book.

Easy access to the benchmark codes and results is one of the most important aspects of the Parkbench effort. Therefore the Methodology chapter also describes the performance database and its graphical front-end which can now be accessed by ftp and via the World-Wide Web. We cover this in Chapter-5 of this book. The Methodology chapter of the Parkbench report also gives procedures for carrying out benchmarking and for optimisation.

The low-level benchmarks chapter of the report describes a set of synthetic tests designed to measure the basic properties of the computer hardware as seen by the user through the Fortran software. They could therefore be described as "architectural" benchmarks. These benchmarks measure the wall-clock execution time of simple DO-loops as a function of loop length, and express the results in terms of two-parameters, namely $(r_\infty, n_{\frac{1}{2}})$. They measure the arithmetic rate of a single processor (or node) for a selection of different loop kernels, the communication latency and rate between different nodes, the ratio or balance between arithmetic and communication, and the time for synchronisation. The correct measurement of time is essential to any benchmark measurement, and two benchmarks are included to test the validity of the computer clock that is used. It is important that this clock measures wall-clock time with sufficient precision. The $(r_\infty, n_{\frac{1}{2}})$ parameters are fully explained in Chapter-3 of this book.

The kernels-benchmark chapter of the Parkbench report describes a set of subroutine-level benchmarks such as one would find in a scientific subroutine library. These include matrix manipulation routines from the University of Tennessee, Fourier transforms, and the solution of partial differential equations by multi-grid and finite-difference methods. The existing NAS Parallel benchmarks [10, 5], including the embarrassingly parallel Monte-Carlo, Fourier transform, conjugate gradient and integer sort have been contributed by NASA to this section. It is hoped eventually to explain the performance of the kernel level benchmarks in terms of the basic parameters measured by the low-level benchmarks, but this has not yet been achieved. We refer the reader to the Parkbench report itself for a detailed description of these benchmarks.

Compact applications are the core computational part of real application but stripped of the features that might make the codes proprietary. The committee plans to collect codes from a wide range of application areas, so that users of the performance database should be able to select applications typical of their type of work. The areas to be covered include Climate Modelling and Meteorology, Seismology, Fluid Dynamics, Molecular Dynamics, Plasma Physics, Computational Chemistry, Quantum Chromodynamics, Reservoir Modelling and Finance. The first edition of the Parkbench Report contained no codes in this section, but gave a procedure for submitting such codes to the committee for consideration. The content of this chapter is expected to grow gradually over the years in future editions, however, the intention is to keep the number of benchmarks to a minimum consistent with the purpose of covering the most important application areas.

The last chapter of the report on compiler benchmarks was included in response to the interest in the proposed High Performance Fortran (HPF) language [37] as a possible common interface to the many different parallel computer architectures and programming models currently existing. A key issue affecting the viability of HPF is the run-time efficiency of the execution code produced by HPF compilers. This chapter defines a standard set of HPF code fragments for testing HPF compilers. This topic is not covered in this book.

1.3 Other Benchmarking Activities

The first step in the assessment of the performance of a parallel computer system is to measure the performance of a single logical processor of the multi-processor system. There exist already many good and well-established benchmarks for this purpose, notably the Linpack benchmarks and the Livermore Loops. These are not part of the Parkbench suite of programs, but Parkbench recommends that these be used to measure single-processor performance, in addition to some specific low-level measurements of its own which are described in Chapter-3. We now describe these other benchmarks and benchmarking activities with their advantages and disadvantages. A very good review of other benchmarks is given by Weicker [78, 79].

1.3.1 Livermore Loops

The Livermore loops (more correctly called the Livermore Fortran Kernels, or LFK) are a set of 24 (originally 14) Fortran DO-loops extracted from operational codes used at the Lawrence

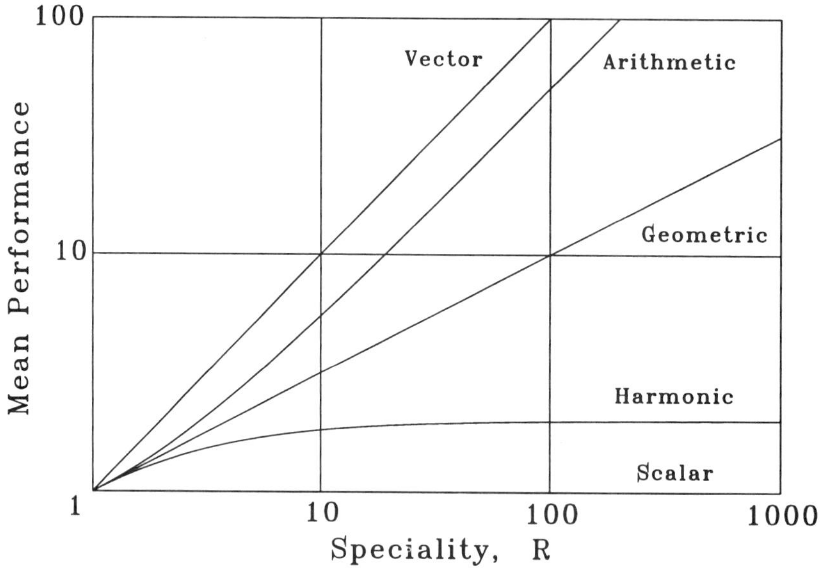

Figure 1.1: Graph showing the effects of different types of averaging of a scalar and a vector benchmark. R is the ratio of the maximum (vector) to minimum (scalar) computing rates, or speciality of the computer.

Livermore National Laboratory. The benchmark, written and distributed by Frank McMahon [58, 59] has been used since the early 1970s to assess the floating-point arithmetic performance of computers and compilers. It originated the use of millions of floating-point operations per second (Mflop/s) as the unit for expressing arithmetic performance. The loops are a mixture of vectorisable and non-vectorisable loops, and test rather fully the computational capabilities of the hardware and the skill of the software in compiling efficient optimised code and in vectorisation. A separate Mflop/s performance figure is given for each loop, and various mean values are computed (arithmetic, harmonic, geometric, median, 1st and 3rd quartiles). The loops are timed for a number of different loop lengths, and weighted averages can be formed in an attempt to represent the benchmarker's particular workload [59].

Although it is tempting to reduce the multiple performance figures produced by this benchmark to a single value by averaging, we think that this is a rather meaningless diversion. It can be easily shown that the arithmetic mean of the individual loop performance rates, corresponds to spending the same time executing each kernel, and that this average tends to be influenced most strongly by the vectorised loops. The harmonic mean, on the other hand, corresponds to performing the same quantity of arithmetic (Mflop) in each loop, and this can be shown to be influenced most by the performance on the worst non-vectorisable loops. The geometric mean does not seem to have any obvious interpretation, although McMahon states that it represents the properties of the Livermore workload quite well. However it is the least biased of the averages and is, for this reason, probably the best to use if a single figure must be produced. These effects are illustrated in Fig. 1.1 which shows the effect of averaging two loops (a non-vectorisable scalar loop and a vectorisable loop) as a function of the ratio, R, of the vector to the scalar rate. For large R, the arithmetic mean follows closely the rate of the vector loop, whilst the harmonic mean never exceeds twice the scalar rate. The geometric mean, on the other hand, is not biased towards either of the two loops that are averaged, and lies nicely between the two extremes.

Table 1.1: Minimum and maximum performance rates for the Livermore loops, together with the speciality ratio and theoretical peak performance. The performance is also expressed as a percentage of the peak.

Computer	Minimum Mflop/s	Maximum Mflop/s	Speciality Ratio	Theoretical Peak Mflop/s
NEC SX2	4.5	1024	227	1300
	0.3%	80%		
ETA 10G	2.2	488	222	644
	0.3%	76%		
CRAY Y-MP1	2.8	295	105	333
	0.8%	88%		
CRAY-2	2.0	228	114	488
	0.4%	46%		
CRAY X-MP	2.2	207	94	235
	0.9%	88%		
IBM 3090e 180VF	2.2	54	24	
IBM RS/6000 530 25MHz	1.3 2.6%	24 48%	18	50
Stardent GS2025	0.45 0.5%	42 52%	93	80
Stardent Titan	0.21 1.3%	13 81%	62	16
Sun 4/110	0.17	1.2	7	
INMOS T800-20MHz	0.10 6.6%	0.94 62%	10	1.5

1.3.2 Speciality Ratio

In our view the value of the Livermore loops is that they give a good idea of the range of performance that will be obtained in floating-point scientific code, and that to reduce this valuable information to a single number makes no sense at all. It is the distribution of performance that is important, especially the maximum and minimum performance rates, and their ratio. We call the ratio of maximum to minimum performance the *speciality ratio* or simply the *speciality*, because it does measure how limited (or special) is the application area for the computer, and how difficult it is to program (because algorithms must be devised to use loops with a format similar to those that execute near the maximum rate or, in the context of vector computers, vectorise well). Assuming a distribution of problems covering more or less uniformly the range of the Livermore loops, then a computer with a high speciality (e.g. 1000) will only perform near its maximum rate for those loops that are near the top of the Livermore performance distribution. Applications using loops near the low end of the distribution will perform at only 0.1% of the best performance, and these applications would be unsuitable. We would describe such a computer as being rather special purpose, because it only performs near its peak and advertised performance for a limited set of problems. If, however the speciality were nearer to the ideal of unity, then all problems would be computed at close to the maximum advertised rate, and we would have a very general purpose computer.

We show in section 1.4.6 that the speciality of a computer is the product of the number of processors and the vector to scalar processing ratio of a single processor. This is based on the highest performance being the vector rate times the number of processors, and the worst performance being the scalar mode on a single processor. The problem with massively parallel computers is that their speciality is becoming very large, 1000 to 10 000, and their field of application is limited to carefully optimised massively parallel problems. The value of the Livermore Loops is that they provide a measured value for the speciality ratio of a computer, based on the performance of loops found in actual production code.

Table-1.1 gives the minimum and maximum performance and the speciality ratio for a number of supercomputers and workstations, together with the theoretical peak performance. We can see that the speciality ratio varies from about 10 for the scalar computers (Sun4, T800, RS/6000) to several hundred for those with vector architectures (NEC, ETA, Cray, Stardent). The IBM 3090VF has a low speciality ratio for a vector machine, owing to the very modest speedup provided by its vector pipelines, compared with the Cray or NEC. It is believed that it was a deliberate design decision not to make the speciality of the 3090 too high, because a higher speciality also means a higher number of dissatisfied customers (there are more customers with problems performing poorly compared to the peak rate).

1.3.3 Linpack

The original Linpack benchmark is a Fortran program for the solution of 100x100 dense set of linear equations by L/U decomposition using Gauss elimination (we call this Linpack100). It is distributed by J. J. Dongarra [24] of the University of Tennessee and Oak Ridge National Laboratory (formerly at the Argonne National Laboratory) and results for a wide variety of computers from PCs to supercomputers are regularly published, and are available over Internet and the World-Wide Web. The results are quoted in Mflop/s for both single and double precision and no optimisation of the Fortran code is permitted, except for that provided automatically by the compiler. Linpack100 is therefore intended as a spot measurement of the performance obtained on existing unoptimised Fortran code. Because the program is very small, it is a very easy and quick benchmark to run. However, most of the compute time is consumed by a vectorisable AXPY loop (scalar $\times$ vector $+$ vector), the benchmark tests very little than this operation but since there is considerable memory traffic, memory access capabilities are exercised. In order to show the maximum capabilities of the computer the benchmark has been extended to include larger problems of 300x300 and 1000x1000 matrices. In particular the Linpack1000 can be coded and optimised in any way, so that the results from this benchmark can be regarded as representing the maximum performance that is likely to be obtained with optimised code on a highly vectorisable and parallelisable problem. The ratio of the optimised to the original Fortran performance is a measure of the possible gain that is likely to be achieved by optimisation. Table-1.2 shows that for the vector computers listed, this gain is typically of the order of ten. The latest Linpack results are available over the World-Wide Web (WWW) at URL address:

`http://www.netlib.org/benchmark/to-get-lp-benchmark`

1.3.4 The Perfect Club

The Perfect Club [16] benchmarking group was setup in 1987 by David Kuck (U. Illinois), Joanne Martin (IBM) and others, in order to provide a set of portable complete application codes (initially in Fortran). It was felt that the earlier benchmarks were too simple to give a proper measurement of a computers capability (or lack of it). The initial release of the

Table 1.2: Results in Mflop/s for the Linpack benchmark, for an unoptimised small problem in Fortran, and a large optimised problem, compared with the theoretical peak performance. The performance is also expressed as a percentage of the peak.

Computer	Fortran n=100	Optimised n=1000	Ratio	Theoretical Peak Mflop/s
CRAY Y-MP 8 CPU 6ns	200 7%	2144 80%	11	2667
CRAY X-MP 4 CPU 8.5ns	149 16%	822 87%	6	940
ETA 10G 1 CPU 7ns	93 14%	496 77%	5	644
CRAY-2S 4 CPU 4.1ns	82 4%	1406 72%	17	1951
CRAY-2 4 CPU 4.1ns	62 3%	1406 72%	23	1951
NEC SX2 1 CPU 6ns	43 3%	885 68%	20	1300
Convex C240 4 CPU 40ns	27 13%	166 83%	6	200
Fujitsu VP400 1 CPU 7ns	20 2%	521 46%	26	1142
CDC Cyber 205 4 pipe 20ns	17 4%	195 48%	11	400
IBM 3090 180sVF 1 CPU 15ns	16 12%	92 69%	6	133

benchmark suite comprised 13 codes with typically 10 000 or more lines of Fortran. Although the distributed code must be run as a baseline measurement, optimisation is permitted provided a diary is kept of the time taken and the improvement obtained. The advantage of this benchmark is that it does test the computer on substantial real problems. The disadvantage is that, because of the complexity of the programs, it is not usually possible to interpret and analyse the results in terms of the computer design. The activities of the Perfect Club have now be merged with that of the SPEC group which is described next.

1.3.5 SPEC Benchmarks

The Systems Performance Evaluation Cooperative (SPEC [73]) is an industrial initiative started in 1989 by Apollo, HP, MIPS and SUN amongst others, with the objective of standardising the benchmarking of new computers, which it was felt had become very confused. The results are published in the Electrical Engineering Times. As with the Perfect Club, only complete application codes are considered, and the first release contained 10 Fortran and C codes, including the well-known Spice circuit simulation program. The interpretation of the results, however is different. The VAX 11/780 has been adopted as a reference computer, and the SPEC ratio for each benchmark program is the ratio of the time to execute on the VAX divided by the time to execute on the computer being tested. This relative performance figure is obtained for each of the programs, and the geometric mean of the ten ratios is quoted as the SPEC Mark for the computer. Because of the relative and averaged nature of this number, it is very difficult to relate it to the basic hardware characteristics of the computer. A more detailed discussion of the SPEC benchmarks, including a brief description of each benchmark, is given by Dixit [21, 22].

Although initially founded by manufacturers to assess the performance of single processor workstations, SPEC has now extended its remit to include the new generation of high performance multiprocessors under the aegis of the SPEC-hpc group. In 1994/5 discussions were held between SPEC-hpc and Parkbench in order to coordinate their activities, and Parkbench whilst remaining a separate organisation is now a member of the SPEC-hpc group. A common code-development policy is currently under discussion for the joint preparation of application benchmark codes (thus excluding Parkbench low-level and kernel benchmarks). This will recognise Parkbench 's policy of free and open access to the source of all codes and results, and the SPEC-hpc policy of controlled access.

1.3.6 Euroben

This European initiative was launched in 1989 by DoDuc, Friedli, Gentzsch, Hockney and Steen [29] based primarily on the existing benchmarks of van der Steen. It is intended as an integrated benchmark suite spanning the range of complexity from simple kernels to complete applications. It comprises the following modules:

1. Kernels (basic arithmetic operations, interpreted with $(r_\infty, n_{\frac{1}{2}})$)
2. Basic algorithms (e.g. Linear equations, FFT)
3. Small applications (e.g. ODEs, PDEs, I/O)
4. Substantial real world applications

Since its foundation Euoben has held informal workshops once every year where those working on benchmark projects could interact and discuss their work. These have proved to be very useful meetings for those working in the field. The initial benchmark set was designed to assess single scalar and vector processors, but with the advent of parallel message-passing computers, a fifth module has been added to the above by incorporating the Genesis Benchmarks which are described next.

1.3.7 Genesis Distributed-Memory Benchmarks

All the benchmarks above were developed for testing the performance of a single CPU, and although they may make use of the vector facilities of a computer by automatic compiler vectorisation, and the multiple CPUs or nodes by automatic compiler parallelisation, none of the above benchmarks are written specifically to make efficient use of the latest generation of highly-parallel distributed-memory computers with communication based on message passing between nodes. The 'Genesis' distributed-memory benchmarks were setup to fill this gap. They arose from the collaborative European 'Genesis' project for the design and construction of a distributed memory supercomputer. They have been used to evaluate the performance of the German Suprenum computer, and the design proposals for its successor. Their organisation and distribution as a benchmark suite is coordinated under the leadership of Professor Tony Hey [36] of Southampton University, UK.

The Genesis benchmarks comprise synthetic code fragments, application kernels, and full application codes. The first version of the benchmark had seven codes including FFT, PDE, QCD, MD, and equation solving. Each code had a standard Fortran77 serial reference version and a distributed memory version using either the `SEND/RECEIVE` statements in the Suprenum extension of the Fortran language or the PARMACS `SEND/RECEIVE` communication macros which are automatically translated to calls the native communication subroutines of the target computer by the PARMACS preprocessor. In most cases there is also a timing computation/communication analysis which should enable the measured times to be related to the hardware computation and communication rates. As well as the Suprenum, the codes have been run on the shared-memory Isis, Cray X-MP, ETA-10, and Alliant, and on the distributed-memory NCube, Parsys SN1000 Supernode, Intel iPSC2, iPSC/860 and Ametek. The benchmarks and results are reported in a series of papers published in *Concurrency: Practice and Experience* [1, 35], and in [36, 2]. With the founding of the Parkbench group, most of the Genesis benchmarks have been incorporated into the low-level and kernel sections of Parkbench , and there will be no further separate development of the Genesis benchmarks. The latest Genesis and Parkbench results can be viewed graphically on WWW using the Southampton GBIS (see section-5.5.6)

1.3.8 RAPS Benchmarks

The RAPS Consultative Forum for Computer Manufacturers was founded in 1991 as part of a proposal for European Esprit III funding by Geerd Hoffmann of the European Centre for Medium Range Weather Forecasts (ECMWF). Standing for Real Applications on Parallel Systems, RAPS brought together the leading weather centres and nuclear laboratories in Europe with the manufacturers of the new generation of MPPs and experienced benchmarking groups from the universities (Southampton and Utrecht), in order to assess the suitability of these computers for their types of large simulation problem. Although the application for European funding was unsuccessful, the group has remained active on a self-financing basis. Industrial members pay a substantial fee to take part in the group, and generally the codes and results are confidential to the group. Because of the large purchasing power of the large laboratories and weather centres most major computer vendors take part, and regular workshops are held.

1.3.9 NAS Parallel Benchmarks (NASPB)

An entirely different type of benchmark was proposed in 1991 by the Numerical Aeronautical Simulation (NAS) group at NASA Ames Laboratory in California, under the initiative primarily of David Bailey and Horst Simon. Conscious of the fact that highly parallel computers differed widely in their detailed architecture and therefore required widely different coding strategies to obtain optimum performance, they proposed that benchmarks should only be defined algorithmically and that no actual parallel benchmark code should be provided. The computer manufacturers were then at liberty to write their own optimum parallel benchmark code in whatever way they wished, provided the published rules describing the algorithm were obeyed. A serial Fortran single-processor code is, however, provided as an aid to implementors and to clarify the algorithmic definition.

Initially the algorithmic descriptions were specified in a NASA report [10] and then published in the International Journal for Supercomputer Applications [5]. Results and comparisons between different MPP computers have also been appeared in NASA reports [9]. The latest version of the rules [6] and results [8] are available from the NAS World-Wide Web Server. David Bailey was a founder member of the Parkbench group, and many of the NAS benchmarks are now also distributed as part of Parkbench .

The above 'paper-and-pencil' benchmarks are known as the NAS Parallel Benchmarks (NASPB) and comprise eight algorithms extracted from important NAS codes and identified by two-letter codes. Five of these are 'kernel' algorithms, and three are simulated CFD applications. The kernels are an embarrassingly parallel (EP) Monte-Carlo algorithm that requires virtually no communication, a multigrid (MG) algorithm for the solution of a 3D Poisson equation, a conjugate gradient (CG) algorithm for eigen-value determination, a 3D FFT used in the solution of partial differential equations by spectral methods, and an integer sorting (IS) algorithm that is used in particle-in-cell (PIC) codes. The best parallel performance and scaling is naturally seen with the EP benchmark, and the worst with the MG and CG benchmarks which contain difficult communication. The IS benchmark is unique in this set as it contains no floating-point operations. The simulated CFD applications use three different methods for solving the CFD equations taken from three different types of NASA simulations: symmetric successive over-relaxation (SSOR) to solve lower-upper diagonal (LU) equations, the solution of a set of scalar pentadiagonal (SP) equations, and the solution of a set of block tridiagonal (BT) equations. A short description of the algorithms and some results is to be found in [7]. Some results of the NASPB are shown in Chapter 5 in Fig. 5.18 (MG) and Fig. 5.19 (LU). The latest NASPB results are available on the WWW at URL address:

`http://www.nas.nasa.gov/NAS/NPB/`

1.4 Usefulness of Benchmarking

Evaluating the performance of a Massively Parallel Processor (MPP) is an extremely complex matter, and it is unrealistic to expect any set of benchmarks to tell the whole story. In fact, the performance of MPPs is often so variable that many people have justifiable questioned the usefulness of benchmarking, and asked whether it is worth while at all. We believe, however, that benchmarks do give very useful information about the capabilities of such computers and that, with appropriate caveats, benchmarking is worth while. The activities of the Parkbench group suggest that this opinion is quite widely held.

In truth we should admit that there is not yet enough data or experience to judge the value of benchmarking, and some benchmark tests are clearly more valuable than others. It is important also to realise that benchmark results do not just reflect the properties of the computer alone, but are also testing how well the benchmark/implementation matches the computer architecture, and how well the algorithm is coded.

1.4.1 The Holy Grail of Parallel Computing

The definition of a single number to quantify the performance of a parallel computer has been the Holy Grail of parallel computing, the unsuccessful search for which has been a distraction from the understanding of parallel performance for years. Aad van der Steen has expresses this succinctly when he states [75]:

> "*The* performance of an MPP is a *non-existing* object."

The reason for this is that parallel performance can vary by such large amounts that the quoting of a single number, which must of necessity be some average of widely varying results, is pretty meaningless. The search for the best definition of such a number is therefore pointless and distracts from the real quest, which is to understand this variation of performance as a function of the computing environment, the size of the problem, and the number of processors. This implies finding timing formulae which match the widely different results with reasonable accuracy, and relating the constants in these to a small number of parameters describing the computer. This is the approach that is taken in Chapter-4 of this book under the title "Computational Similarity and Scaling".

Aad van der Steen calls this approach "benchmarking for knowledge" which he characterises as benchmarking which:

1. Tries to relate system properties with performance
2. Tries to inter-relate performance results
3. Tries to relate performance results of applications with lower-level benchmark results

This has the following advantages:

1. Yields insight into the systems of interest
2. Results might be generalised (with care)
3. Systems may be compared
4. May yield useful extrapolations

Parkbench and Euroben [29] are examples of benchmarking initiatives that have taken the above approach.

1.4.2 Limitations and Pitfalls

From his extensive experience, Aad van der Steen [75] has also summarised the limitations and pitfalls of computer benchmarking as follows:

1. Benchmarks cannot answer questions that are not asked
2. Specific application benchmarks will *not* tell you (or very little) about the performance of other applications without a proper analysis of these applications
3. General benchmarks will not tell you all details of the performance of your specific application (but may help a good deal in understanding it).
4. To be equipped to understand correctly the story that a benchmark tells us, one should know the background.

1.4.3 Only Certain Statement

The performance of MPPs is so dependent on the details of the particular problem being solved, the particular MPP on which it is executed, and the particular software interface being used, that the only certain statement that can be made is something like:

This particular *implementation* of
.... *benchmark* of
.... *size* on
.... *number of processors*

executes in

this particular *time* on
.... *computer* using
.... *compiler* with
.... *optimisation level* and
.... *communication library*

No generalisation from the above very specific statement can be made without extensive studies of the scaling properties of the particular benchmark/computer combination that is being considered. Generally insufficient data is available and this is *not* done, but the study of performance scaling with the problem size and number of processors is an active area of current research and some recent work is reported in Chapter-4.

It is a truism to say that the best benchmark is the program that one is actually going to run on the computer, and there is one case in which knowledge of the performance of one program is all that is required. That is the case of a computer installation dedicated to the execution of one program, for example a Weather Centre, or a computer dedicated to automotive crash simulation. In this case the only benchmark required is the particular program that is used.

Even in these installations, however, there is likely to be an element of program development and the need for a computer that performs well over a range of different applications. For such general-purpose use, benchmarking is much less reliable, and the usual approach is to provide a range of test problems which hopefully cover the likely uses at the installation. Thus the Parkbench Compact Applications are to be chosen from a range of applications.

1.4.4 Most Useful Benchmarks (Low-level)

In many respects the low-level architectural benchmarks are the most useful because their results measure the basic capabilities of the computer hardware, and in principle the performance of higher level benchmarks should be predictable in terms of them. They are certainly the easiest to analyse, interpret and understand. Because they are simple in concept, they are relatively easy to implement efficiently and sensibly, and they also are a good test of the capabilities of a compiler. They are also easy to run and take very little computer time, no more than a few minutes for each benchmark.

In their recent paper with the revealing title "A careful interpretation of simple kernel benchmarks yields the essential information about a parallel supercomputer", Schönauer and Häfner [69] also subscribe to this view, and state in their abstract:

> "We believe that simple kernel benchmarks are better for the measurement of parallel computers than real-life codes because the latter may reveal more about the quality of one's software than about the properties of the measured computer."

In this they also recognise that low-level benchmarks are the one's to use to measure the fundamental properties of the computer architecture.

1.4.5 Least Useful Benchmarks (Application)

In some sense application benchmarks are the least useful because it is very difficult to make generalisations from the results if the benchmark is not exactly what the user proposes to run. Such a generalisation would require a timing model for the benchmark which has been validated against measurements for several problem sizes (say 4) and with sufficient values for 'p' (the numbers of processors used) for each size to adequately show the p-variation. This means about 50 measurements of a code that might take, typically 5 to 15 minutes to run. Clearly such an exercise is time and resource consuming, and indeed is a research project in its own right. Furthermore the timing model may not prove to be accurate enough for confidence (although an accuracy of 20 to 30 percent might be considered good enough). Nevertheless the fitting of such timing models has been conducted with some success by the "Genesis" benchmarking group at Southampton University [1].

Given a such validated timing model which depends on problem size, number of processors, and a small number of basic computer hardware parameters (e.g. scalar arithmetic rate, communication bandwidth and latency), then limited extrapolation of the benchmark performance can be made with some confidence to other problem sizes and number of processors than were used in the benchmark measurements.

1.4.6 Best/Worst Performance Ratio

When considering MPPs for general purpose computing one has to consider the range of performance of the MPP from its worst to its best. The ratio of the best to the worst performance has previously been called both the "instability" or the "speciality" of the MPP. If this ratio is large then there is a greater premium attached to tuning the code, and a greater penalty for writing bad code. The performance may therefore vary greatly for apparently minor code changes, in short the system is unstable to code changes. Equally there is only going to be a relatively small set of codes that perform well, and one would say the system was therefore rather specialised.

In order to understand this issue better, consider an MPP with p scalar/vector processors with:

$$\text{scalar rate} = r^s_\infty \tag{1.1}$$

$$\text{asymptotic vector rate} = r^v_\infty \tag{1.2}$$

where, typically, the ratio of asymptotic vector to scalar rate is

$$\frac{r^v_\infty}{r^s_\infty} = 5 \tag{1.3}$$

The *best* performance will be obtained when all processors perform at the vector rate:

$$r_{\text{best}} = p * r^v_\infty \tag{1.4}$$

and the *worst* performance is obtained when only one processor executes serial code at the scalar rate:

$$r_{\text{worst}} = r^s_\infty \tag{1.5}$$

Hence the instability/speciality is

$$\frac{r_{\text{best}}}{r_{\text{worst}}} = p * \frac{r^v_\infty}{r^s_\infty} \tag{1.6}$$

We note first that the speciality is proportional to the number of processors, so that the more highly parallel the MPP, the more difficult it is to use efficiently, and the more variable is its performance.

For the traditional parallel vector computer such as the Cray-C90 with say 2-16 processors and a vector to scalar rate ratio of 5, we have a speciality between 10 and 80. However for MPPs where 'p' might start from a few 100 and rise to a few thousand, we find the speciality is much larger and in the range from a few 100 to several thousand.

This higher value for the instability/speciality is the fundamental problem with MPPs, and shows that one does not want parallelism for its own sake, but only if there is no other way of obtaining the performance required.

In many problems that are not ideally suited for parallel processing, the performance is limited by the time taken for the essentially serial code that must be executed on a single processor (or is, for convenience, alternatively, repeated on every processor on the MPP). In either case, the time taken is the same as if the serial code were executed on only one processor of the MPP. This serial code gives rise to the Amdahl limit [3] on parallel performance, which says that even if the rest of the code is parallelised efficiently and takes zero time, the execution cannot take less than the time to execute the serial code (see section 4.1). The only way to make the Amdahl performance limit less severe is to select an MPP with faster processors. This selection can be made by examining the results of simple single-processor benchmarks such as RINF1 from Parkbench low-level benchmarks, and there is no need to perform any parallel benchmarks.

The way to judge how well different MPPs cope with the problem of serial code is to compare the worst-performance obtained over the whole range of benchmarks, and to choose the MPP with the best worst-performance, and this leads to the following principle of choice.

1.4.7 Hockney's Principle of Parallel-Computer Choice

Because of the great variability (or instability) of performance of MPPs on different problems, the following principle is suggested for selecting such computers:

1. Of *all computers* that
   (a) Have *acceptable* performance on your problems
   (b) Are *within budget*
2. The *wisest choice* is the computer with
   (a) The *best worst-Performance*
3. Which *usually* means the computer with
   (a) The *least* Parallelism
   (b) The *highest single-processor* performance

We call this the *Principle of Minimal Parallelism*. Some readers may find the above counter intuitive and contrary to the currently popular opinion that: "in parallel computing, the more processors the better". However a number of other researchers have recently expressed the same opinion. For example, David Bailey writes in *Parallel Computing Research* [12]:

> There has been considerable convergence in the hardware area lately. For example, it is now generally acknowledged that the best parallel systems are those that have a moderate number of high-powered nodes as opposed to tens of thousands of weak nodes. However this is 20/20 hindsight"

In a similar vain, Aad van der Steen [75] gives the following rules of thumb for benchmarking:

1. A small amount of fast processors (for your application) is always preferable over many slow processors to do the same job.

2. No communication is always faster than fast communication, however fast.

1.4.8 Conclusion

Benchmarks can be used to measure the main characteristics of MPPs. In particular attention must be made not only to the best performance, but also to the worst performance obtained on each MPP. For special problems and carefully tuned code the best performance may be the most appropriate performance indicator, but for general-purpose use the worst performance is more likely to be more important. This means that the highest performance processors are desirable and the least parallelism. Hence we conclude: MPP Yes, but for general purpose use, 'M' should *not* stand for Massively, but rather for Minimally, Modestly or Moderately Parallel Processors. This is a recognition of the fact that there is an overhead in going parallel that is proportional to the number of processors, and that this overhead should be kept to a minimum (see section 3.3.4).

Chapter 2

Methodology

One of the aims of the Parkbench committee was to set standards for benchmarking methodology and result reporting. The methodology chapter of the Parkbench report begins this process by defining a set of units, symbols and metrics for expressing benchmark results. This section of the report was written primarily by David Bailey and Roger Hockney, and is reproduced here, followed in section 2.7 by an example of the use of the different performance metrics.

2.1 Objectives

One might ask why anyone should care about developing a standardised, rigorous and scientifically tenable methodology for studying the performance of high-performance computer systems. There are several reasons why this is an important undertaking:

1. To establish and maintain high standards of honesty and integrity in our profession.
2. To improve the status of supercomputer performance analysis as a rigorous scientific discipline.
3. To reduce confusion in the high-performance computing literature.
4. To increase understanding of these systems, both at a low-level hardware or software level and at a high-level, total system performance level.
5. To assist the purchasers of high-performance computing equipment in selecting systems best suited to their needs.
6. To reduce the amount of time and resources vendors must expend in implementing multiple, redundant benchmarks.
7. To provide valuable feedback to vendors on bottlenecks that can be alleviated in future products.

It is important to note that researchers in many scientific disciplines have found it necessary to establish and refine standards for performing experiments and reporting the results. Many scientists have learned the importance of standard terminology and notation. Chemists, physicists and biologists long ago discovered the importance of "controls" in their experiments. The issue of repeatability proved crucial in the recent "cold fusion" episode. Medical researchers have found it necessary to perform "double-blind" experiments in their field. Psychologists and sociologists have developed highly refined experimental methodologies and advanced data analysis techniques. Political scientists have found that subtle differences in the phrasing of a question can affect the results of a poll. Researchers in many fields have found that environmental factors in their experiments can significantly influence the measured results; thus they must carefully report all such factors in their papers.

If supercomputer performance analysis and benchmarking is ever to be taken seriously as a scientific discipline, certainly its practitioners should be expected to adhere to standards that prevail in other disciplines. This document is dedicated to promoting these standards in our field.

2.2 Units and Symbols

A rational set of units and symbols is essential for any numerate science including benchmarking. The following extension of the internationally agreed SI system of physical units and symbols [65, 14] is made to accommodate the needs of computer benchmarking.

The value of a variable comprises a pure number stating the number of units which equal the value of the variable, followed by a unit symbol specifying the unit in which the variable is being measured. A new unit is required whenever a quantity of a new nature arises, such as e.g. the first appearance of vector operations, or message sends. Generally speaking a unit symbol should be as short as possible, consistent with being easily recognised and not already used. The following have been found necessary in the characterisation of computer and benchmark performance in science and engineering. No doubt more will have to be defined as benchmarking enters new areas.

New unit symbols and their meaning:

1. flop: floating-point operation
2. inst: instruction of any kind
3. intop: integer operation
4. vecop: vector operation
5. send: message send operation
6. iter: iteration of loop
7. mref: memory reference (read or write)
8. barr: barrier operation
9. b: binary digit (bit)
10. B: byte (groups of 8 bits)
11. sol: solution or single execution of benchmark
12. w: computer word. Symbol is lower case (W means watt)
13. tstep: timestep

When required a subscript may be used to show the number of bits involved in the unit. For example: a 32-bit floating-point operation flop_{32}, a 64-bit word w_{64}, also we have $\mathrm{b} = \mathrm{w}_1$, $\mathrm{B} = \mathrm{w}_8$, $\mathrm{w}_{64} = 8\mathrm{B}$.

Note that flop, mref and other multi-letter symbols are inseparable four or five-letter symbols. The character case is significant in all unit symbols so that e.g. Flop, Mref, W_{64} are incorrect. Unit symbols should always be printed in roman type, to contrast with variables names which are printed in italic. Because 's' is the SI unit for seconds, unit symbols like 'sheep' do not take 's' in the plural. Thus one counts: one flop, two flop, ..., one hundred flop etc. This is especially important when the unit symbol is used in ordinary text as a useful abbreviation, as often, quite sensibly, it is.

SI provides the standard prefixes:

1. k : kilo meaning 10^3
2. M : mega meaning 10^6
3. G : giga meaning 10^9
4. T : tera meaning 10^{12}

This means that we cannot use M to mean 1024^2 (the binary mega) as is often done in describing computer memory capacity, e.g. 256 MB. We can however introduce the new Parkbench prefixes:

1. K : meaning binary kilo i.e. 1024, then use a subscript 2 to indicate the other binary versions
2. M_2 : binary mega i.e. 1024^2
3. G_2 : binary giga i.e. 1024^3
4. T_2 : binary tera i.e. 1024^4

In most cases the difference between the mega and the binary mega (4%) is probably unimportant, but it is important to be unambiguous. In this way one can continue with existing practice if the difference doesn't matter, and have an agreed method of being more exact when necessary. For example, the above memory capacity was probably intended to mean $256\mathrm{M}_2\mathrm{B}$.

As a consequence of the above, an amount of computational work involving 4.5×10^{12} floating-point operations is correctly written as 4.5 Tflop. Note that the unit symbol Tflop is never pluralised with an added 's', and it is therefore incorrect to write the above as 4.5 Tflops which could be confused with a rate per second. The most frequently used unit of performance, millions of floating-point operations per second is correctly written Mflop/s, in analogy to km/s. The slash or solidus is necessary and means 'per', because the 'p' is an integral part of the unit symbol 'flop' and cannot also be used to mean 'per'. Mflop/s can also be written Mflops^{-1} which is the recommended SI procedure, however the use of the slash is allowed in SI and seems more natural in this context.

2.3 Time Measurement

Before other issues can be considered, we must discuss the measurement of run time. In recent years a consensus has been reached among many scientists in the field that the most relevant measure of run time is actual wall-clock elapsed time. This measure of time will be required for all Parkbench results that are posted to the database.

Elapsed wall-clock time means the time that would be measured on an external clock that records the time-of-day or even Greenwich mean time (GMT), between the start and finish of the benchmark. We are not concerned with the origin of the time measurement, since we are taking a difference, but it is important that the time measured would be the same as that given by a difference between two measurements of GMT, if it were possible to make them. It is important to be clear about this, because many computer clocks (e.g. Sun Unix function `ETIME`) measure elapsed CPU time, which is the total time that the process or job which calls it has been executing in the CPU. Such a clock does not record time (i.e. it stops ticking) when the job is swapped out of the CPU. It does not record, therefore, any wait time which must be included if we are to assess correctly the performance of a parallel program. On some systems, scientists have found that even for programs that perform no explicit I/O, considerable "system" time is nonetheless involved, for example in fetching certain library routines or other data.

Only timings actually measured may be cited for Parkbench benchmarks (and we strongly recommend this practice for other benchmarks as well). Extrapolations and projections, for instance to a larger number of nodes, may not be employed for any reason. Also, in the interests of repeatability it is highly recommended that timing runs be repeated, several times if possible.

Two low-level benchmarks are provided in the Parkbench suite to test the precision and accuracy of the clock that is to be used in the benchmarking. These should be run first, before any benchmark measurements are made. They are:

1. TICK1 - measures the precision of the clock by measuring the time interval between ticks of the clock. A clock is said to tick when it changes its value.

2. TICK2 - measures the accuracy of the clock by comparing a given time interval measured by an external wall-clock (the benchmarker's wrist watch is adequate) with the same interval measured by the computer clock. This tests the scale factor used to convert computer clock ticks to seconds, and immediately detects if a CPU-clock is incorrectly being used.

The fundamental measurement made in any benchmark is the elapsed wall-clock time to complete some specified task. All other performance figures are derived from this basic timing measurement. The benchmark time, $T(N;p)$, will be a function of the problem size, N, and the number of processors, p. Here, the problem size is represented by the vector variable, N, which stands for a set of parameters characterising the size of the problem: e.g. the number of mesh points in each dimension, and the number of particles in a particle-mesh simulation. Benchmark problems of different sizes can be created by multiplying all the size parameters by suitable powers of a single scale factor, thereby increasing the spatial and particle resolution in a sensible way, and reducing the size parameters to a single size factor (usually called α).

We believe that it is most important to regard execution time and performance as a function of at least the two variables (N,p), which define a parameter plane. Much confusion has arisen in the past by attempts to treat performance as a function of a single variable, by taking a particular path through this plane, and not stating what path is taken. Many different paths may be taken, and hence many different conclusions can be drawn. It is important, therefore, always to define the path through the performance plane, or better as we do here, to study the shape of the two-dimensional performance hill. In some cases there may even be an optimum path up this hill.

2.4 Floating-Point Operation Count

Although we discourage the use of millions of floating-point operations per second as a performance metric, it can be a useful measure if the number of floating-point operations, $F(N)$, needed to solve the benchmark problem is carefully defined.

For simple problems (e.g. matrix multiply) it is sufficient to use a theoretical value for the floating-point operation count (in this case $2n^3$ flop, for nxn matrices) obtained by inspection of the code or consideration of the arithmetic in the algorithm. For more complex problems containing data-dependent conditional statements, an empirical method may have to be used. The sequential version of the benchmark code defines the problem and the algorithm to be used to solve it. Counters can be inserted into this code or a hardware monitor used to count the number of floating-point operations. The latter is the procedure followed by the Perfect Club (see section-1.3.4). In either case a decision has to be made regarding the number of flop that are to be credited for different types of floating-point operations, and we see no good reason to deviate from those chosen by McMahon [59] when the Mflop/s measure was originally defined. These are:

add, subtract, multiply	1 flop
divide, square-root	4 flop
exponential, sine etc.	8 flop (this figure will be adjusted)
`IF(X .REL. Y)`	1 flop

Some members of the committee felt that these numbers, derived in the 1970s, no longer correctly reflected the situation on current computers. However, since these numbers are only used to calculate a nominal benchmark flop-count, it is not so important that they be accurate. The important thing is that they do not change, otherwise all previous flop-counts would have to be renormalised. In any case, it is not possible for a single set of ratios to be valid for all computers and library software. The committee agreed that the above ratios should be kept for the time being, but that the value for the transcendental function was unrealistic and would be adjusted later after research into a more realistic and higher value.

We distinguish two types of operation count. The first is the nominal benchmark floating-point operation count, $F_B(N)$, which is found in the above way from the defining Fortran77 sequential code. The other is the actual number of floating-point operations performed by the hardware when executing the distributed multi-node version, $F_H(N,p)$, which may be greater than the nominal benchmark count, due to the distributed version performing redundant arithmetic operations. Because of this, the hardware flop count may also depend on the number of processors on which the benchmark is run, as shown in its argument list.

2.5 Performance Metrics

The conclusions drawn from a benchmark study of computer performance depend not only on the basic timing results obtained, but also on the way these are interpreted and converted into performance figures. The choice of the performance metric, may itself influence the conclusions. For example, do we want the computer that generates the most megaflop per second (or has the highest Speedup), or the computer that solves the problem in the least time? It is now well known that high values of the first metrics do not necessarily imply the second property. This confusion can be avoided by choosing a more suitable metric that reflects solution time directly, for example either the Temporal, Simulation or Benchmark performance, defined below. This issue of the sensible choice of performance metric is becoming increasing important with the advent of massively parallel computers which have the potential of very high megaflop rates, but have much more limited potential for reducing solution time.

Given the time of execution $T(N;p)$ and the flop-count $F(N)$ several different performance measures can be defined. Each metric has its own uses, and gives different information about the computer and algorithm used in the benchmark. It is important therefore to distinguish the metrics with different names, symbols and units, and to understand clearly the difference between them. Much confusion and wasted work can arise from optimising a benchmark with respect to an inappropriate metric. The principal performance metrics are the Temporal, Simulation, Benchmark and Hardware performance. The objections to the use of Speedup and Efficiency are then discussed.

2.5.1 Temporal Performance

If we are interested in comparing the performance of different algorithms for the solution of the same problem, then the correct performance metric to use is the *Temporal Performance*, R_T, which is defined as the inverse of the execution time

$$R_T(N;p) = T^{-1}(N;p) \tag{2.1}$$

The units of Temporal performance are, in general, solutions per second (sol/s), or some more appropriate absolute unit such as timesteps per second (tstep/s). With this metric we can be sure that the algorithm with the highest performance executes in the least time, and is therefore the best algorithm. We note that the number of flop does not appear in this definition, because the objective of algorithm design is not to perform the most arithmetic per second, but rather it is to solve a given problem in the least time, regardless of the amount of arithmetic involved. For this reason the Temporal performance is also the metric that a computer user should employ to select the best algorithm to solve his problem, because his objective is also to solve the problem in the least time, and he does not care how much arithmetic is done to achieve this.

2.5.2 Simulation Performance

A special case of Temporal performance occurs for simulation programs in which the benchmark problem is defined as the simulation of a certain period of physical time, rather than a certain number of timesteps. In this case we speak of the *Simulation Performance* and use units such as *simulated days per day* (written sim-d/d or 'd'/d) in weather forecasting, where the apostrophe is used to indicate 'simulated'; or *simulated pico-seconds per second* (written sim-ps/s or 'ps'/s) in electronic device simulation. It is important to use Simulation performance rather than timestep per second if one is comparing different simulation algorithms which may require different sizes of timestep for the same accuracy (for example an implicit scheme that can use a large timestep, compared with an explicit scheme that requires a much smaller step). In order to maintain numerical stability, explicit schemes also require the use of a smaller timestep as the spatial grid is made finer. For such schemes the Simulation performance falls off dramatically as the problem size is increased by introducing more mesh points in order to refine the spatial resolution: the doubling of the number of mesh-points in each of three dimensions can reduce the Simulation performance by a factor near 16 because the timestep must also be approximately halved. Even though the larger problem will generate more Megaflop per second, in forecasting, it is the simulated days per day (i.e. the Simulation performance) and not the Mflop/s, that matter to the user.

As we see below, benchmark performance is also measured in terms of the amount of arithmetic performed per second or Mflop/s. However it is important to realise that it is incorrect to compare the Mflop/s achieved by two algorithms and to conclude that the algorithm with the highest Mflop/s rating is the best algorithm. This is because the two algorithms may be performing quite different amounts of arithmetic during the solution of the same problem. The Temporal performance metric, R_T, defined above, has been introduced to overcome this problem, and to provide a measure that can be used to compare different algorithms for solving the same problem. However, it should be remembered that the Temporal performance only has the same meaning within the confines of a fixed problem, and no meaning can be attached to a comparison of the Temporal performance on one problem with the Temporal performance on another.

2.5.3 Benchmark Performance

In order to compare the performance of a computer on one benchmark with its performance on another, account must be taken of the different amounts of work (measured in flop) that the different problems require for their solution. Using the flop-count for the benchmark, $F_B(N)$, we can define the *Benchmark Performance* as

$$R_B(N;p) = F_B(N)/T(N;p) \tag{2.2}$$

The units of Benchmark performance are Mflop/s ($<$ *benchmark name* $>$), where we include the name of the benchmark in parentheses to emphasise that the performance may depend strongly on the problem being solved, and to emphasise that the values are based on the nominal benchmark flop-count. In other contexts such performance figures would probably be quoted as examples of the so-called *sustained* performance of a computer. We feel that the use of this term is meaningless unless the problem being solved and the degree of code optimisation is quoted, because the performance is so varied across different benchmarks and different levels of optimisation. Hence we favour the quotation of a selection of Benchmark performance figures, rather than a single sustained performance, because the latter implies that the quoted performance is maintained over all problems.

Note also that the flop-count $F_B(N)$ is that for the defining sequential version of the benchmark, and that the same count is used to calculate R_B for the distributed-memory (DM) version of the program, even though the DM version may actually perform a different number of operations. It is usual for DM programs to perform more arithmetic than the defining sequential version, because often numbers are recomputed on the nodes in order to save communicating their values from a master processor. However such calculations are redundant (they have already been performed on the master) and it would be incorrect to credit them to the flop-count of the distributed program.

Using the sequential flop-count in the calculation of the DM programs Benchmark performance has the additional advantage that it is possible to conclude that, for a given benchmark, the implementation that has the highest Benchmark performance is the best because it executes in the least time. This would not necessarily be the case if a different $F_B(N)$ were used for different implementations of the benchmark. For example, the use of a better algorithm which obtains the solution with less than $F_B(N)$ operations will show up as higher Benchmark performance. For this reason it should cause no surprise if the Benchmark performance occasionally exceeds the maximum possible Hardware performance. To this extent Benchmark performance Mflop/s must be understood to be nominal values, and not necessarily exactly the number of operations executed per second by the hardware, which is the subject of the next metric. The purpose of Benchmark performance is to compare different implementations and algorithms on different computers for the solution of the same problem, on the basis that the best performance means the least execution time. For this to be true $F_B(N)$ must be kept the same for all implementations and algorithms.

2.5.4 Hardware Performance

If we wish to compare the observed performance with the theoretical capabilities of the computer hardware, we must compute the actual number of floating-point operations performed, $F_H(N;p)$, and from it the actual *Hardware Performance*

$$R_H(N;p) = F_H(N,p)/T(N;p) \tag{2.3}$$

The Hardware performance also has the units Mflop/s, and will have the same value as the Benchmark performance for the sequential version of the benchmark. However, the Hardware performance may be higher than the Benchmark performance for the distributed version, because the Hardware performance gives credit for redundant arithmetic operations, whereas the Benchmark performance does not. Because the Hardware performance measures the actual floating-point operations performed per second, unlike the Benchmark performance, it can never exceed the theoretical peak performance of the computer.

Assuming a computer with multiple-CPUs each with multiple arithmetic pipelines, delivering a maximum of one flop per clock period, the theoretical peak value of hardware performance is

$$r^* = \frac{fl.pt.pipes/CPU}{clock.period} \times number.CPUs \tag{2.4}$$

with units of Mflop/s if the clock period is expressed in microseconds. By comparing the measured hardware performance, $R_H(N;p)$, with the theoretical peak performance, we can assess the fraction of the available performance that is being realised by a particular implementation of the benchmark.

2.6 What's Wrong with Speedup?

It has been common practice for a long time in the parallel computing community, to use the Speedup of an algorithm or benchmark as a single figure-of-merit (that Holy Grail again) for judging both computer and algorithmic performance, as though it were an absolute measure of performance. A little thought shows that this practice is, in general, invalid and can lead to very false conclusions. Computers have even been purchased on the basis of high Speedup numbers, by people forgetting to take into account the slow speed of their processors. It is also common experience that you can always obtain good scaling behaviour or speedup if one uses slow enough processors, and that obtaining good speedup with fast processors is difficult. In other words to look at Speedup in isolation without taking into account the speed of the processors is unrealistic and pointless. For this reason the Parkbench committee came out strongly against the use of the Speedup metric in reporting benchmark results, as is seen in the following paragraph taken from the report. This is followed by a paragraph (not in the report) discussing the use of Speedup to compare algorithms. Interestingly, Speedup does appear in the theory of scaling presented in Chapter-4 as the natural and correct way to define "dimensionless performance". The fact that the word "dimensionless" is used reinforces the fact that Speedup is *not* an absolute measure, and cannot correctly be used as one.

2.6.1 Speedup and Efficiency

"Parallel speedup" is a popular metric that has been used for many years in the study of parallel computer performance. However, its definition is open to ambiguity and misuse because it always begs the question "speedup over what?"

Speedup is usually defined as

$$\frac{T_1}{T_p} \tag{2.5}$$

where T_p is the p-processor time to perform some benchmark, and T_1 is the one-processor time. There is no doubt about the meaning of T_p — this is the measured time $T(N;p)$ to perform the benchmark. There is often considerable dispute over the meaning of T_1: should it be the time for the parallel code running on one processor, which probably contains unnecessary parallel overhead, or should it be the best serial code (possibly using a different algorithm) running on one processor? Many scientists feel the latter is a more responsible choice, but this requires research to determine the best practical serial algorithm for the given application. If at a later time a better algorithm is found, current speedup figures might be considered obsolete. An additional difficulty with this definition is that even if a meaning for T_1 is agreed to, there may be insufficient memory on a single node to store an entire large problem. Thus in many cases it may be impossible to measure T_1 using this definition.

One principal objective in the field of performance analysis is to compare the performance of different computers by benchmarking. It is generally agreed that the best performance corresponds to the least wall-clock execution time. In order to adapt the speedup statistic for benchmarking, it is thus necessary to define a single reference value of T_1 to be used for all calculations. It does not matter how T_1 is defined, or what its value is, only that the same value of T_1 is used to calculate all speedup values used in the comparison.

However, defining T_1 as a reference time unrelated to the parallel computer being benchmarked unfortunately has the consequence that many properties that many people regard as essential to the concept of parallel speedup are lost:

1. It is no longer necessarily true that the speedup of the parallel code on one processor is unity. It may be, but only by chance.

2. It is no longer true that the maximum speedup using p-processors is p.

3. Because of the last item, efficiency figures computed as speedup divided by p are no longer a meaningful measure of processor utilisation.

There are other difficulties with this formulation of speedup. If we use T_1 as the run time on a very fast single processor (currently, say, a Cray T90 or a NEC SX-3), then manufacturers of highly parallel systems will be reluctant to quote the speedup of their system in the above way. For example, if the speedup of a 100 processor parallel system over a single node of the same system is a respectable factor of 80, it is likely that the speedup computed from the "standard" T_1 would be reduced to 10 or less. This is because a fast vector processor is typically between five and ten times faster than the RISC processors used in many highly parallel systems of a comparable generation.

Thus it appears that if one sharpens the definition of speedup to make it an acceptable metric for comparing the performance of different computers, one has to throw away the main properties that have made the concept of speedup useful in the past.

Accordingly, the Parkbench committee decided the following:

1. No speedup statistic will be kept in the Parkbench database.

2. Speedup statistics based on Parkbench benchmarks must never be used as figures-of-merit when comparing the performance of different systems. We further recommend that speedup figures based on other benchmarks not be used as figures of merit in such comparisons.

3. Speedup statistics may be used in a study of the performance characteristics of an individual parallel system. But the basis for the determination of T_1 must be clearly and explicitly stated.

4. The value of T_1 should be based on an efficient uniprocessor implementation. Code for message passing, synchronisation, etc. should not be present. The author should also make a reasonable effort to insure that the algorithm used in the uniprocessor implementation is the best practical serial algorithm for this purpose.

5. Given that a large problem frequently does not fit on a single node, it is permissible to cite speedup statistics based on the timing of a smaller number of nodes. In other words, it is permissible to compute speedup as T_m/T_p, for some m, $1 < m < p$. If this is done, however, this usage must be clearly stated, and full details of the basis of this calculation must be presented. As above, care must be taken to ensure that the unit timing T_m is based on an efficient implementation of appropriate algorithms.

2.6.2 Comparison of Algorithms

If an algorithm solves a problem in a time, T, then the most unambiguous definition of algorithmic performance is the Temporal performance $R_T = T^{-1}$, that is to say the number of solutions per second. If two algorithms are compared with this absolute definition of performance, there is no doubt that the algorithm with the highest performance executes in the least time. If, however, we use the Speedup of an algorithm (which is a relative measure) as the definition of performance – as is very frequently done – then we have to be very careful, because it is by no means always true that the algorithm with the greatest Speedup executes in the least time, and false conclusions can easily be drawn.

With the definition of Speedup as in Eqn.(2.5) we can make the following observations:

1. Speedup is performance arbitrarily scaled to be unity for one processor

2. Speedup is performance measured in arbitrary units that will differ from algorithm to algorithm if T_1 changes.

3. Speedup cannot be used to compare the relative performance of two algorithms, unless T_1 is the same for both.

4. The program with the worst Speedup may execute in the least time, and therefore be the best algorithm.

By taking the ratio of two performances, the concept of Speedup is throwing away all knowledge of the absolute performance of an algorithm. It is a number without units. Thus if we compare the speed of two algorithms by comparing their Speedups, it is like comparing the numerical values of the speeds of two cars when the speed of one is measured in m.p.h. and the speed of the other in cm/s. No meaning can be attached to such a comparison because the units are different. Comparisons of Speedups are only valid if the unit of measurement is the same for both, that is to say T_1 is the same for both. An example of phenomenon (4) is given by Cvetanovic et.al [20] when they found that an SOR algorithm for the solution of Poisson's equation had a much better Speedup than an ADI algorithm, although ADI was the best algorithm because it executed in the least time.

The problems with the use of Speedup to compare algorithms lie entirely with the definition and measurement of T_1. For example, for large problems which fill a massively parallel distributed system, it is almost certainly impossible to fit the whole problem into the memory of a single processor, so that T_1, in fact, may be impossible to measure. There is also the problem as to what algorithm to time on one processor. Is it the parallel algorithm run on one processor, or should it more fairly be the best serial single processor algorithm, which is almost certain to be different? If the latter choice is taken, it begs more questions than it answers, because there may be disputes about what is the best serial algorithm, and whether it has been programmed and optimised with the same care as the parallel algorithm which was the main object of the research (rather unlikely). Also if in time a better serial algorithm is invented, then in principle all previous Speedup curves should be rescaled.

The point that we make in this section is that all the problems raised in the last paragraph are completely spurious and unimportant. If we measure performance in absolute terms using one of the metrics defined above, none of the above questions and confusions arise.

2.6.3 Speedup Conclusions

We can summarise the above discussion as follows:

1. Speedup can be used
   (a) to study, in isolation, the scaling of one algorithm or benchmark on one computer.
   (b) as a dimensionless variable in the theory of scaling (see Chapter-4)
2. But it *must not* be used to compare
   (c) different algorithms or benchmarks on the same computer, or
   (d) the same algorithm or benchmark on different computers,

because T_1 (the unit of measurement) may change. One should therefore use absolute (inverse time) units for (c) and (d), such as

1. Temporal Performance in sol/s or tstep/s
2. Simulation Performance in sim-d/d or sim-ps/s
3. Benchmark (nominal) Performance in Mflop/s(< *benchmark name* >)

because these show not only the scaling behaviour but also the absolute computing rate.

2.7 Example of the LPM1 Benchmark

The use of the above metrics will be illustrated by the case of the LPM1 benchmark from the "Genesis" benchmark suite [36, 1, 2]. Although these results are rather old [47, 48] and involve computers that are no longer of current interest, they have been extensively analysed and the references show the same results expressed in all the metrics defined above, demonstrating rather well their relative merits.

2.7.1 LPM1 Benchmark

The LPM1 (standing for Local Particle-Mesh) benchmark is the time-dependent simulation of an electronic device using a particle-mesh or PIC-type algorithm. It has a two-dimensional (r, z) geometry and the standard problem size has the fields computed on a regular mesh (33x75). The fields computed are E_r, E_z, and B_ϕ (into the mesh). The electron distribution in the device is represented by a group of 500 to 1000 particles simulating electron clouds, each of which has its positions and velocity stored. The timestep proceeds as follows:

1. The particle coordinates are inspected, and the current of each particle is assigned to the mesh points in its neighbourhood.

2. The fields are advanced for a short timestep according to Maxwell's equations.

3. Each particle is accelerated according to Newton's Laws with the new field values obtained by inspecting the mesh values near each particle.

4. Injection takes place along the cathode of the device according to the value of the normal E-field.

This is called a *local* particle-mesh simulation because the timescale is such that only neighbouring (i.e. local) field values are required during the update of the field equations. In contrast a global simulation might require the solution of, say, Poisson's equation over the whole mesh. Because of this locality of data, a geometric subdivision of the problem between the nodes of a transputer network (or other distributed-memory computer) is natural. The existing distributed-memory implementation takes this approach, and performs a one-dimension domain decomposition. For p processors, the device is divided into p approximately equal slices in the z-direction, and each slice is assigned to a different processor. Each processor is responsible for computing the fields, and moving the particles in its region of space. When particles move out of this region, their coordinates are transferred to the correct neighbouring processor, which computes their subsequent motion. The timestep is such that a particle will move no further than to a neighbouring processor in one timestep. The processors are therefore configured as a chain, and during the timestep loop, communication is only required between neighbouring processors in the chain. The communication comprises an exchange of the edge values of the fields and currents, together with the coordinates of any particles that are moving between processors.

The standard benchmark is defined as the simulation of one nanosecond of device time using a (33,75) mesh and starting from an empty device. During this time electrons are emitted from the cathode starting at the left of the device, and the electron cloud gradually fills the device from left to right. After one nanosecond the device is about half filled with electrons. Larger problem sizes are introduced by using more mesh-points in the z-direction, and computing on a $(33, 75\alpha)$ mesh, where α is the problem-size factor (=1,2,4,8). It is found empirically that the total number of particles in the system at the end of one nanosecond grows somewhat faster than proportionality to α and is represented within a few percent by

$$N_{end} = 628\alpha^{1.17} \tag{2.6}$$

In order to produce a timing formulae we must make a simple model of the filling of the device. Since the number of mesh-points computed by each processor is approximately the same, the timing of the multiprocessor simulation will be determined by the time take by the processor containing the most particles. Since the device is filled from left to right this will be the left most processor in the chain, and will be called the critical processor. We assume that the critical processor fills at a constant rate until it contains $2N_{end}/p$ particles (which are its share of the particles present at the end of the simulation), and that the number of particles in the critical processor remains constant at this value for the rest of the benchmark run. The '2' in the above expression arises because only the first $p/2$ processors in the chain contain any particles at the end of the benchmark (i.e. the device only half fills). The average number of particles in the critical processor is then given by

$$N_1 = \frac{2N_{end}}{p}(1 - 1/p) \tag{2.7}$$

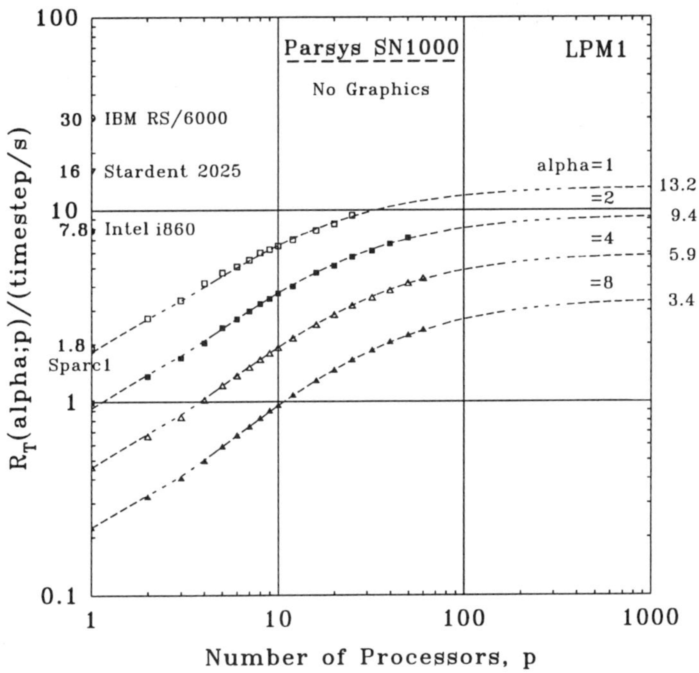

Figure 2.1: Temporal performance of the LPM1 benchmark for four problem sizes as a function of the number of processors used, on a Parsys SN1000 Supernode with 25 MHz T800 transputers. α is the problem-size factor.

where the first term would apply alone if the critical processor was filled with particles from the beginning, and the second term is the correction taking into account that there are fewer particles during the filling process.

An inspection of the Fortran code shows that, per timestep, there are 46 flop per mesh point and 116 flop per particle in the most important terms of the operation-count. This leads us to define the nominal flop-count per timestep for the benchmark to be

$$F_B(\alpha) = 46 \times 33 \times 75\alpha + 116 \times N_{end}/2 \tag{2.8}$$

where the last term assumes the number of particles increases linearly from zero to N_{end} during the reference single-processor run.

2.7.2 Comparison of Performance Metrics

Figure 2.1 shows the Temporal performance in units of timestep per second as a function of the number of processors for four problem sizes on a 64-processor Parsys SN1000 Supernode. Each node comprises a 25 MHz Inmos T800 transputer with 4 MByte of local memory. The problem size is increased by multiplying the number of mesh points in the z-direction by a factor $\alpha = 1, 2, 4, 8$. For simplicity there is no change to the number of mesh points in the r-direction, nor is there domain decomposition in the r-direction. That is to say that each processor computes all the r-points in its region of the z-direction.

The time of execution of the benchmark has been fitted within a few percent to the formula

$$T(\alpha;p) = a + b\alpha + \frac{c\alpha}{p}\left\{1 + 1.28\alpha^{0.17}(1 - 1/p)\right\} \tag{2.9}$$

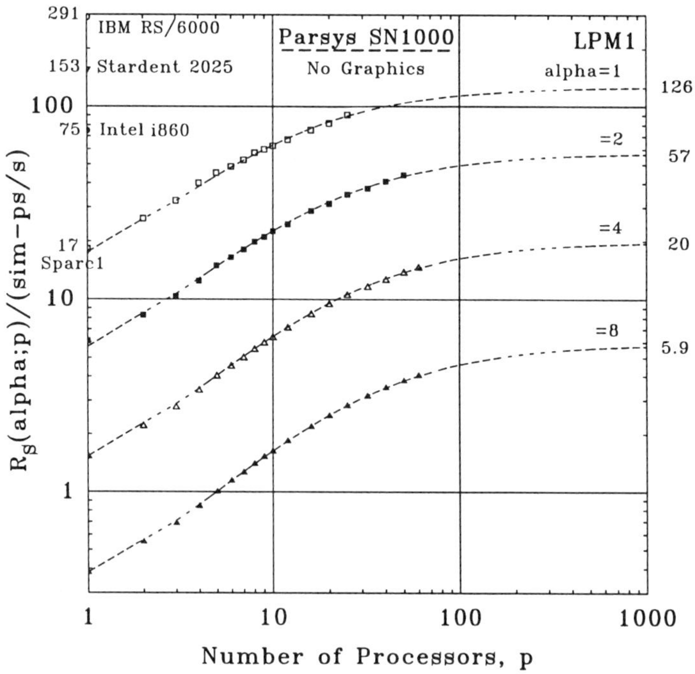

Figure 2.2: Simulation performance of the LPM1 benchmark for four problem sizes as a function of the number of processors used, on a Parsys SN1000 Supernode with 25 MHz T800 transputers. α is the problem-size factor. The timing data used is the same as Fig.2.1

where $a = 0.0453$s, $b = 0.0307$s, and $c = 0.358$s. The first two terms containing a and b take account of the non-parallelised part of the code and arise from both communication and arithmetic. They determine the execution time and corresponding performance that in theory would arise in the limit as the number of processors p goes to infinity (The so-called Amdahl saturation limit). The last term which is proportional to c and inversely proportional to the number of processors p, is from the parallelised part of the code which is subdivided between the processors. The two parts of the parallelised code are represented by the two terms within the braces. The first term (=1) comes from the field calculation on the mesh points. The complicated second term comes from an empirical fit to the number of particles in the critical processor that contains the most particles, and therefore determines the calculation time. The values of a, b, c are obtained by a least-squares fit to the Temporal performance. The theoretical performance curves derived from Eqn.(2.9) are shown as dotted lines in all the figures, whilst the measured data points are plotted as unjoined symbols.

Because Fig. 2.1 uses absolute performance units, it can also show on the y-axis the performance of a number of single-processor workstations. The best of these is the IBM RS/6000-530(25MHz) with a performance of 30 tstep/s, followed by the Stardent 2025 at 16 tstep/s, and the Intel iPSC/860 (using the Portland Group Fortran Compiler). It is clear that the Amdahl limit prevents the parallel Parsys SN1000 code from exceeding about 13 tstep/s, however many processors are used, and that this code runs slower than a single processor IBM RS/6000 or Stardent 2025.

Figures 2.2 and 2.3 show the same performance data as figure 2.1 but plotted as Simulation performance and Benchmark Mflop/s(LPM1) performance. Both the Temporal and

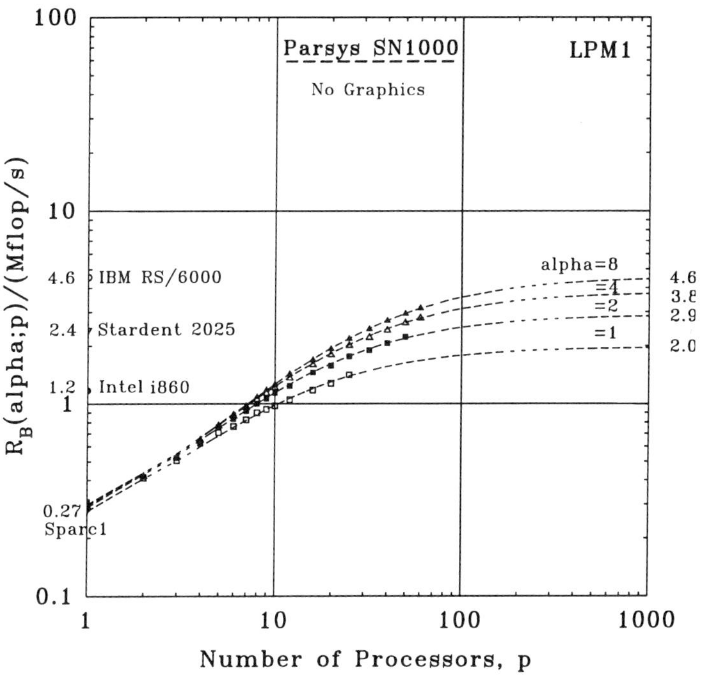

Figure 2.3: Benchmark performance of the LPM1 benchmark for four problem sizes as a function of the number of processors used, on a Parsys SN1000 Supernode with 25 MHz T800 transputers. α is the problem-size factor. The timing data used is the same as Fig.2.1

Simulation performance show that the smallest problem ($\alpha = 1$) is the fastest (i.e. executes in the least time), and that the fall-off of performance as the problem size increases is much faster for Simulation performance than it is for Temporal performance, because the timestep size has to be decreased as the number of mesh points increases, in order to satisfy the CFL stability criterion.

However the Benchmark Mflop/s is seen to increase with problem size, showing that this metric can be misleading, and does not indicate the relative execution time of different sized simulations. Thus one finds that larger problems generate more Mflop/s, but one should not forget that they take longer to run. Therefore the best strategy for winning the Gordon-Bell Prize (which is based on the highest value of Mflop/s) is to solve the largest problem possible, but the best strategy for solving the problem most quickly is to choose the smallest problem that has the required accuracy. This will generate less Mflop/s but will get the solution in the least time and therefore most economically.

In contrast to the Temporal performance, values of Benchmark performance for all problem sizes on one processor are approximately the same. This is to be expected and simply shows that the formula, $F_B(N)$, for the nominal benchmark flop-count is successful in accounting for the increased work as the the problem size, N, increases. This results in a fan-like appearance for the family of performance curves (one for each N value), and confines them to a relatively small area of a log/log performance graph of $R_B(N;p)$ versus p. This is an advantage when plotting the performance of many computers on one graph in order to compare their performance, because the performance fans for the different computers are less likely to overlap than if the same results were plotted as Temporal performance.

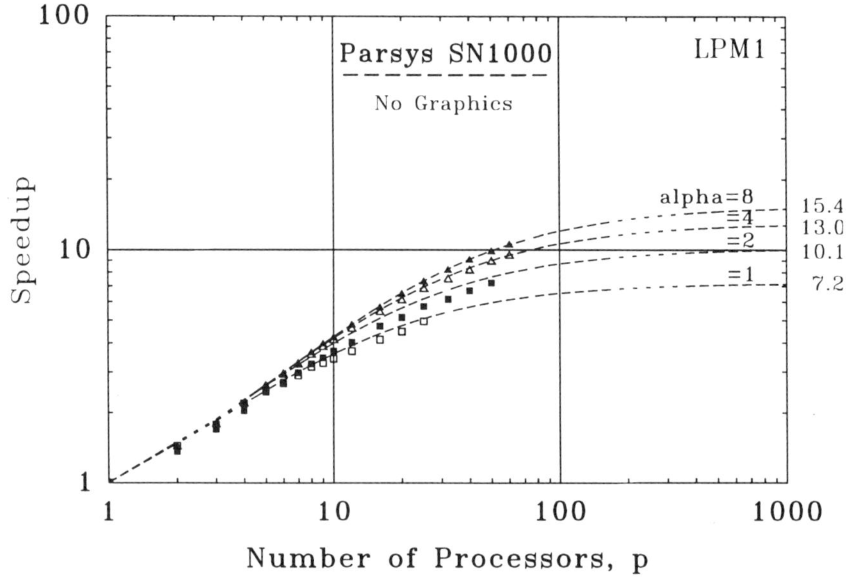

Figure 2.4: LPM1 benchmark results presented in terms of Speedup. The timing data used is the same as Fig.2.1

Figure 2.4 shows the performance results plotted in terms Speedup. The Speedup curves have the same fan-like appearances as those of the Benchmark performance in Mflop/s but we have now lost all knowledge of the absolute value of the performance, and can no longer compare the performance with the other computers. The Speedup, like the Benchmark performance, credits the largest problem ($\alpha = 8$) with the highest performance, whereas we know from the Temporal performance that the largest problem actually takes longer to run.

Figure 2.5 shows the same results plotted in terms of the Efficiency of calculation, that is to say the Speedup divided by the number of processors. Again we have no comparison with other computers, and the largest problem is found to have the highest efficiency although it has the lowest Temporal performance. We also find that the fastest problem (in sim-ps/s) has the lowest efficiency, so that it does not seem that efficiency is measuring what the user wants to know. This example clearly demonstrates why we do not think that benchmark results should be presented in terms of Speedup or Efficiency.

2.8 LPM3 Benchmark

The above LPM1 benchmark was the result of parallelising an existing single-processor workstation simulation code which was not originally designed with parallel processing in mind. As a consequence, the increase of performance as the number of processors increases is not as good as one would like, but it may be quite typical of some other codes parallelised in similar circumstances. Furthermore, during the first 100 timesteps that form the benchmark test, the simulated device is filling up with electrons from the left, and there is a large load imbalance between processors with a large number of particles on the left and those with none or few on the right. This, together with the fact that LPM1 is a two-dimensional program only parallelised along one dimension, leads to very poor scaling behaviour, and limits the number of processors that can be effectively used to a few tens. Nevertheless all these circumstances

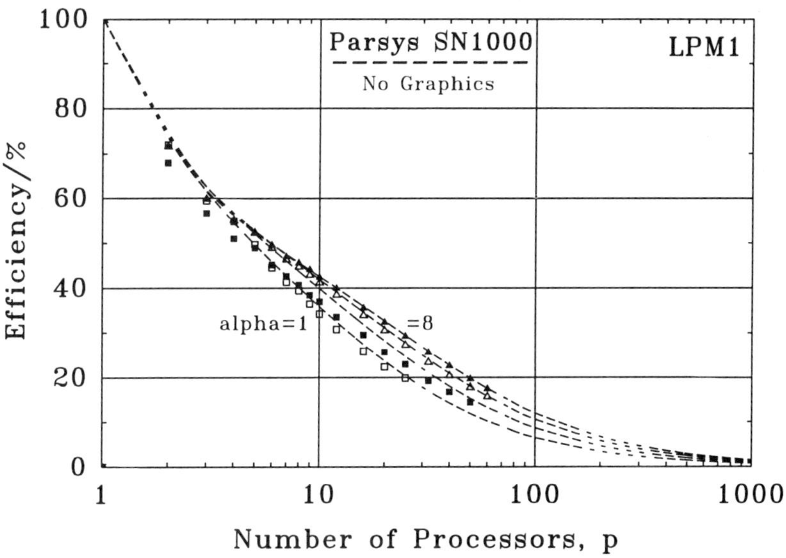

Figure 2.5: LPM1 benchmark results presented in terms of Efficiency. The timing data used is the same as Fig.2.1

are likely to apply to some other real codes when they are being used to simulate real devices with complex geometries, and a good parallel performance on the LPM1 benchmark may be taken as an indication of good performance in unfavourable circumstances.

At the other extreme, the LPM3 benchmark is a version of a code designed from the start to take advantage of parallel processing, and good parallel performance on this code probably indicates the best that can be obtained from parallel processing. First, the simulation is of a large three-dimensional problem which is parallelised in three dimensions, thus giving the potential to use thousands of processors. Secondly the problem solved is chosen to allow for perfect load balancing if a suitable number of processors is used. On the other hand the problem solved can fairly be described as being artificial, and unlike LPM1 it does not represent the simulation of a real electronic device. Hence with LPM3 we expect to get nearly perfect linear scaling, and significant deviations from this indicates a serious problem with either the parallel hardware or software or both.

The LPM3 benchmark code simulates a triply-periodic three-dimensional uniform electron plasma, and is fully described in Eastwood, Arter, Brealey and Hockney [27]. The plasma space is divided into blocks, each of which contains 512 particles representing the electrons, and 64 elements on which the fields are calculated. From the point of view of load balancing on the parallel computer, the block is the smallest unit that can be allocated amongst the processors. The problem size is measured by the number of blocks N_b, and four problem sizes have been used with $N_b = 8, 64, 512, 4096$. These correspond respectively to numbers of particles $N_p = 4\text{K}, 32\text{K}, 256\text{K}, 2\text{M}_2$ where $\text{K} = 1024$ and $\text{M}_2 = \text{K}^2$. The timestep is such that about ten percent of the particles leave each block and enter neighbouring blocks during a timestep. A run of 100 timesteps is chosen as the benchmark test because this can be done in a few minutes for problem sizes and number of processors of interest (tstep/s in the range 0.1 to 10), and the conservation of total number of particles is used as a validity check.

There are two different versions of the benchmark code, which are selected by an input variable. The per-patch version sends a separate message for every patch in the system, and there are 9 patches for every block. In the per-processor version, on the other hand, the patch messages are assembled and sorted in a buffer so that only one message is sent to every other processor to which a given processor is attached. The per-processor code may send 10 or 100 times fewer messages than the per-patch version, and should be significantly faster than the per-patch version on computers with a high message startup time or latency. For computers with low latency there will be little difference between the two versions.

As a demonstration of the use of standard benchmarks to compare the scaling behaviour of large MPPs, the LPM3 benchmark has been run on the Intel Paragon at the Sandia National Laboratory, Albuquerque, with up to 1840 processors, and on the IBM SP2 at the Maui High Performance Computer Center (MHPCC) with up to 128 processors. These results have been obtained under the auspices of AEA Technology, the USAF and the above computer centres.

2.8.1 IBM SP2 Communication Alternatives

We start by using LPM3 to compare the performance of some of the different communication interfaces that are available on the IBM SP2:

PVM3ip: Public domain PVM version 3.0 from the Oak Ridge National laboratory with full Internet communication protocol (ip) between processors, using ethernet connections. May be used to connect processors of different types, as in a cluster of heterogeneous workstations. Word format conversion between different types of processor is accommodated. For these reasons PVM3 has a high startup overhead, but it is implemented on all parallel computers of importance. PVM3 code should therefore be highly portable. (Data was taken at MHPCC on a dedicated partition of 64 MByte thin nodes on 4 January 1995).

PVM3sw: Same as above, but using IBM high-speed switch instead of ethernet connections. (Data was taken at MHPCC on a dedicated partition of 64 MByte thin nodes on 4 January 1995).

PVMe: IBM customised version of PVM. This assumes the processors are all the same and within a single SP2. Most of the internet protocol can be omitted and no word format conversion is needed. The IBM high-speed switch is used. The source code is the same as PVM3 and therefore portable. (Data was taken at MHPCC on a dedicated partition of 64 MByte thin nodes on 7 March 1995, using PVMe 1.3.1, xlf 3.2, poe 1.2.1 under AIX 3).

MPL: This is the "native" IBM SP2 Fortran library of communication subroutines. This provides minimum startup overhead in a high-level language. However MPL is unique to IBM, and MPL code is therefore not portable to other computers. The high-speed switch is used. (Data was taken at MHPCC on a dedicated partition of 64 MByte thin nodes on 7 March 1995, using xlf 3.2, poe 1.2.1 under AIX 3).

In order to compare these alternatives we show in Fig. 2.6 the performance of both versions of the LPM3 benchmark on the 8-block case. The symbol plotted identifies the version used, and the type of dotted line identifies the communication interface. The graph plots Temporal performance in timestep per second (tstep/s) against the number of processors used on a log/log scale, in order to show the scaling of performance with the number of processors. The ideal linear scaling, in which the performance is directly proportional to the number of processors, is shown by the dotted line at 45 degrees.

Public domain PVM3 performs badly with this code, with the per-patch version showing no speedup over single-processor performance with either PVM3ip or PVM3sw (lowest pair of

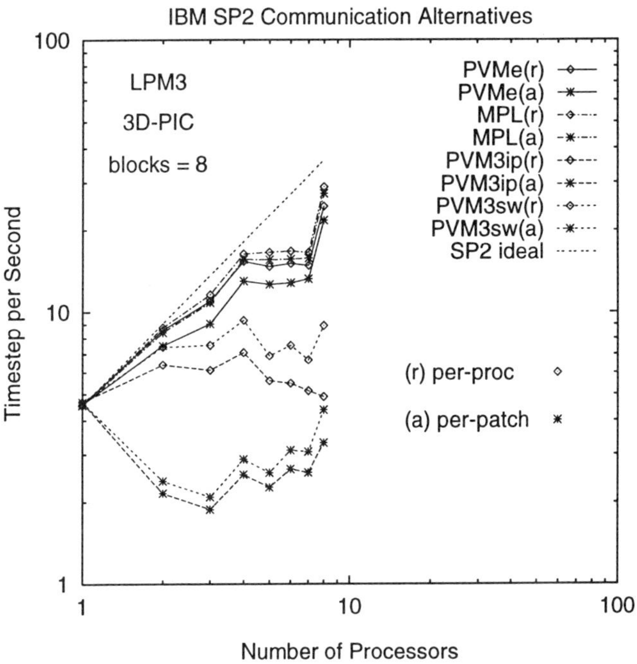

Figure 2.6: Temporal Performance of the LPM3 benchmark measured in units of timestep per second for the 8-block case on the IBM SP2 for four different communication alternatives. Two code versions are shown: (a) per-patch and (r) per-processor. Line type distinguishes the communication interface, and the plotting symbol the code version. (*Data courtesy of AEA Technology and USAF*)

curves). The per-processor version shows a distinct improvement (next higher pair of curves), but the speedup scarcely improves with more than two processors, and is at most two for eight processors (PVM3sw). In both cases use of the switch (sw) is slightly better than ethernet (ip) but the difference is not significant. The advantage of combining the patch messages into a single longer message for each connected processor is clearly seen.

The next higher pair of curves are for IBM PVMe with the per-processor version having the better performance for all processor numbers. The step-like appearance of these curves is due to load imbalance between the processors. For 1, 2, 4 and 8 processors, each processor has the same number of blocks, namely 8, 4, 2 and 1 block, and therefore the same work to do. The load is exactly balanced across the processors, and the performance is optimal. In other cases the load is unbalanced and the performance is determined by the processors with the largest number of blocks. Other processors will have to wait for these critical processors to finish. For 3 processors the largest number of blocks per processor is 3, which explains why the 3 processor performance lies between that for 2 processors (4 blocks per processor) and that for 4-processors (2 blocks per processor). For 5, 6, and 7 processors, there is always at least one processor with two blocks, which explains why the performance for these three cases is almost the same as that for 4 processors (2 blocks per processor). The measured performance for IBM MPL follows the pattern as for PVMe but is in most cases about 10 to 20 percent better. For both PVMe and MPL, there is little difference between the performance of the per-patch and per-processor versions, showing that the message startup time is sufficiently small that reducing the number of messages sent in the per-processor version produces little

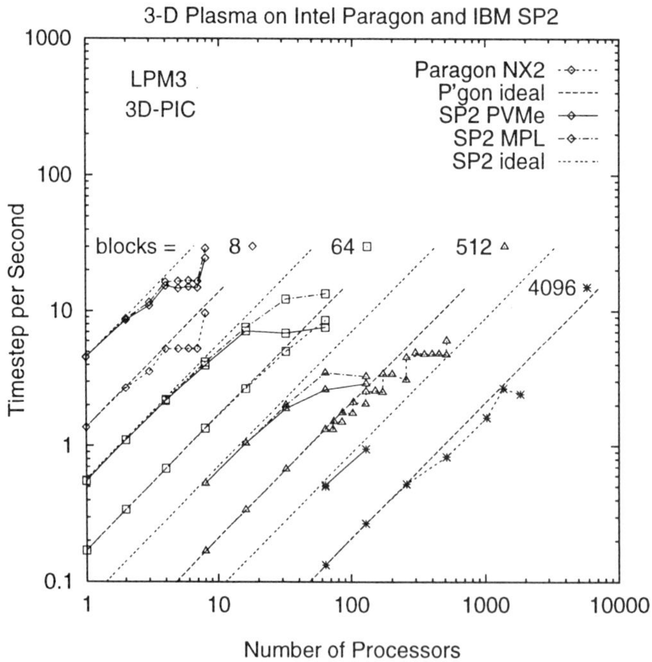

Figure 2.7: Temporal Performance of the LPM3 benchmark (per-processor version) for four problem sizes on the Intel Paragon and IBM SP2. The plotting symbol distinguishes the problem size (number of blocks) and the line type the computer/communication combination. (*Data courtesy of AEA Technology and USAF*)

saving in time.

2.8.2 Comparison of Paragon and SP2

In order to compare the performance of LPM3 on the Intel Paragon with the IBM SP2, we show in Fig. 2.7 the Temporal performance of the per-processor code, which was the best on both computers, for the 4 cases of 8 ,64, 512 and 4096 blocks. The Intel Paragon data is for the NX2 communication library running under the Sandia SUNMOS 1.4.8 operating system which may be considered to be the Intel equivalent of the native IBM MPL interface. Although we have measured both versions on the IBM SP2 under both MPL and PVMe, we plot only the per-processor results which were consistently better than the per-patch results. The plotting symbol used distinguishes the problem size (number of blocks), and the line type the computer and communication combination. The step-like shape of the Intel results is due to load imbalance. Except for the 8-block case the IBM results only show cases of exact load balance, and the underlying step-like behaviour is not seen.

Looking at the performance for one processor we find the IBM SP2 to be about 2.5 times faster than the Intel Paragon, and this difference in single processor performance determines the other results. The dotted lines at 45 degrees show the perfect linear scaling from the one-processor performance for the 8-block case. This assumes that performance is directly proportional to the number of processors and inversely proportional to the number of blocks. The Paragon scales extremely well right up to 1840 processors which is the maximum available, and this enables it to overcome the poorer single-processor performance in the 512 and 4096 block cases, although there is not enough parallelism in the 64 and 8-block cases for this to be possible, because one cannot use more processors than there are blocks. The SP2 performance shows a significant fall off from the ideal scaling for more than 16 processors on the 64-block case and more than 64 processors for the 512 block case. Although there are 400 processors on the Maui SP2, the maximum that can be used routinely in practice is 128, so it has not been possible to test the SP2 scaling beyond 128 processors.

2.8.3 Conclusions

For the best performance on the IBM SP2 it is advisable to use the native communication interface, but portable PVM code run under IBM customised PVMe is almost as good. For the LPM3 benchmark, public domain PVM3 used over either ethernet (corresponding to a cluster of workstations) or the switch has too much startup overhead to give any worth-while speedup. The significantly better single-processor performance of the SP2 compared to the Paragon, gives the SP2 an initial advantage, but this is lost for the larger problems by the better scaling of the Paragon and the much larger number of processor that are available. PVM code has not been run on the Paragon. The per-processor version of the communication interface has consistently better performance than the original per-patch version, but this difference is only significant for PVM3 communications when it is dramatic.

Chapter 3

Low-Level Parameters and Benchmarks

The purpose of the low-level benchmarks is to measure certain important characteristics of the target computer system such as the arithmetic and communication rates and overheads. They are synthetic in the sense that each is designed to measure a particular architectural feature of the computer, and in contrast to the higher-level kernel and application benchmarks, they solve no real problem. They are therefore often aptly referred to as synthetic or as architectural benchmarks.

For the benchmarks to be meaningful, the measured timings must be analysed with respect to simple computational timing models based on a small number of "hardware" performance parameters, which in turn will be used to represent the hardware properties in timing models for the higher-level benchmarks. Thus the definition of the low-level performance parameters and the design of benchmarks to measure them go hand-in-hand, and this chapter is concerned with both aspects of the topic. We use the term hardware here to mean the characteristics of the whole computer system as seen by the user through a high-level language such as Fortran or C, that is to say it not only includes the basic time delays of the hardware circuitry but also includes the delays associated with the software through which it is used. This is an important point because often high values for parameters representing overheads arise not from high hardware circuit and chip delays but from long software overheads. This is particularly true when measuring communication overheads. Fortunately software overheads can usually be reduced more easily than true circuit and chip hardware delays.

We consider first the fundamental measurement of time, and benchmarks to check the precision and value of the computer clock which is to be used in the benchmark measurements (section 3.1); then the definition of theoretical peak performance with which some of the measured performances can be compared, together with the reasons why the performance that is actually realised (often called, rather badly, the "sustained" performance) is usually much less than this peak (section 3.2); then the definition of the $(r_\infty, n_{\frac{1}{2}})$ parameters, an understanding of which is fundamental to the correct interpretation of the low-level benchmarks (section 3.3); then the individual benchmarks are described which assess the arithmetic performance (RINF1, section 3.4), the communication performance (COMMS benchmarks, section 3.5), the balance between arithmetic and communication performance (POLY benchmarks, section 3.6), and the cost of global synchronisation (SYNCH1, section 3.7).

3.1 The Measurement of Time

According to the methodology outlined in Chapter-2 the fundamental measurement in any benchmarking is the measurement of elapsed wall-clock time. Also, because the computer clocks on each processor of a multi-processor parallel computer are not synchronised, it is important that all benchmark time measurements be made with a single clock on one processor of the system. The benchmarks TICK1 and TICK2 have, respectively, been designed to measure the resolution and to check the absolute value of this clock. These benchmarks should be run with satisfactory results before any further benchmark measurements are made.

3.1.1 Timer resolution: TICK1

TICK1 measures the resolution of the clock, which is defined as the time interval between successive ticks of the clock, where a tick is said to occur whenever the clock changes its value. The value of the computer clock is obtained by calling a timer subroutine, and TICK1 makes a succession of calls to this subroutine in a loop which it executes many times. The differences between successive values given by the timer are then examined. If the clock tick is longer than the time taken to enter and leave the timer subroutine, then most of these differences will be zero. When a tick takes place, however, a difference equal to the tick value will be recorded, surrounded by many zero differences. This is the case with clocks of poor resolution: for example most UNIX clocks that tick typically every 10 ms. Such poor UNIX clocks can still be used for low-level benchmark measurements if the benchmark is repeated, say 10 000 times, and the timer calls are made outside this repeat loop (see section 3.4 and Fig. 3.3).

With some computers, such as the CRAY series, the clock ticks every cycle of the computer, that is to say every 2ns on the T90. The resolution of the CRAY clock is therefore five million times better than a poor UNIX workstation clock, and that is quite a difference! If TICK1 is used on such a computer the difference between successive values of the timer is a very accurate measure of how long it takes to execute the instructions of the timer subroutine, and therefore is never zero. TICK1 takes the minimum of all such differences, and all it is possible to say is that the clock tick is less than or equal to this value. Typically this minimum will be several hundreds of clock ticks. With a clock ticking every computer cycle, we can make low-level benchmark measurements without a repeat loop. Such measurements can even be made on a busy timeshared system (where many users are contending for memory access) by taking the minimum time recorded from a sample of, say 10 000 single execution measurements. In this case, the minimum can usually be said to apply to a case when there was no memory access delay caused by other users.

3.1.2 Timer value: TICK2

TICK2 confirms that the absolute values returned by the computer clock are correct, by comparing its measurement of a given time interval with that of an external wall-clock (actually the benchmarker's wristwatch). The time interval is defined by the benchmarker making two key strokes at the computer terminal, one to start and the other to end the interval. A time interval of about two minutes or 100 seconds is recommended, and can easily be measured with an accuracy better than one percent at the keyboard.

Parallel benchmark performance can only be properly measured using the elapsed wall-clock time, because the objective of parallel execution is to reduce this time. Measurements made with a CPU-timer (which only records time when its job is executing in the CPU) are clearly incorrect, because the clock does not record waiting time when the job is out of the CPU. TICK2 will immediately detect the incorrect use of a CPU-time-for-this-job-only clock, because the computer CPU clock will only record about 30ms for what is known to be an elapsed wall-clock time of about a minute. This is because it takes about 15ms to process each keystroke, and the waiting time between the keystokes is not recorded by the CPU timer. An example of a timer that claims to measure elapsed time, and might therefore be thought to be a wall-clock timer, is the returned value of the popular Sun UNIX timer `ETIME`. TICK2 immediately detects that `ETIME` is, in fact, a CPU timer and unsuitable for parallel benchmarking. TICK2 also checks that the correct multiplier is being used in the computer system software to convert clock ticks to true seconds. Since this conversion factor is merely a number built into the subroutine library of the computer, it is quite possible for its value to be inconsistent with the actual frequency of the hardware clock. It is obviously as well to be sure that this multiplier is correct before reporting benchmark performance numbers.

3.2 Peak, Realised and Sustained Performance

Benchmarks are most useful when their results can be related to the true hardware characteristics of the computer. The easiest way to make this connection is to express the result of the benchmark in millions of floating-point operations per second (Mflop/s) and then to compare this number with the theoretical peak Mflop/s, r^*, of the computer hardware, as defined in Eqn.(2.4) which is repeated here for convenience

$$r^* = \frac{fl.pt.pipes/CPU}{clock.period} \times number.CPUs \tag{3.1}$$

Here a unit pipe is defined as one that produces one floating-point result per clock period when full, and by convention we count only add and multiply pipelines. The existence (or not) of division or square root functional units is usually not counted because of their infrequent occurrence, whereas multiply rarely occurs without an accompanying addition. For problems which are dominated by division or square root, then this definition would have to change. The measured Hardware performance (section 2.5.4) with which the peak can be compared is obtained by dividing the number of floating-point operations performed (Mflop) by the time of execution. For low-level benchmarks and kernel/subroutine-level benchmarks such as Linpack the number of floating-point operations performed can be seen by inspection, but for full application codes such as those from the indexPerfect ClubPerfect Club and SPEC a computer hardware monitor may have to be used to measure the number of floating-point operations during the execution of the benchmark.

The theoretical peak performance should be regarded as a kind of 'speed of light' for the computer which could only be reached if the arithmetic units of the computer were busy 100% of the time. Usually this rate can only be achieved when data resides in the local scalar and vector registers, and thus it also can be regarded also as the register-to-register computation rate. The observed performance of benchmarks is degraded by the time taken to move data between the main memory (where the problem data resides) to the registers, and for other organisational operations (usually integer) which do not contribute to the numerical result of the benchmark (e.g. index calculation, DO-loop control tests). For these reasons it is rarely possible to keep the arithmetic units busy all the time doing useful arithmetic, that is to say necessary arithmetic which contributes to the numerical result of the benchmark. In Table-1.1 and Table-1.2 we can see that the realised performance on the Livermore and Linpack benchmarks typically varies from about one percent to about 80 percent of the theoretical peak, depending on the nature of the loop, problem size, and level of optimisation. In these circumstances there is clearly no simple answer to the question "how fast is my computer?" because the answer depends on so many factors.

The theoretical peak performance is the rating usually quoted by a manufacturer, but all that can be said, in truth, is that this is the rate which the manufacturer guarantees his computer cannot exceed. The unrealistic nature of the peak rate has been recognised for some years, and there is a now a tendency for manufacturers also to quote something called a "sustained" performance rate. This term has no precise meaning but seems to be the performance that is actually realised on some real problem, as opposed to a theoretical upper limit. However unless the problem solved is stated this term has no meaning, because as we have seen any computer can have a wide range of performance depending on the problem solved and its size. Some manufacturers quote the Linpack100 as their sustained rate, others quote more favourably a highly optimised FFT code. Clearly so-called sustained rates cannot be compared unless they are for the same problem. The purpose of benchmarking is to define such test problems, and the manufacturers should be asked to quote their performance on a set of benchmarks. In our view, the term sustained performance should be dropped and replaced by the name of a well-defined benchmark whose performance is being quoted. It will then be possible to compare meaningfully the so-called sustained rates of different computers.

We should not, however, despair too much, and the rest of this chapter is devoted to defining simple performance parameters to measure separately different reasons for the degradation of performance from the theoretical peak, and special low-level benchmarks to measure these parameters. The performance parameters should be regarded as properties of the computer hardware and software environment that is being used, and usually go in pairs, the first of which is an asymptotic rate, and the second an overhead or 'sub-half' parameter. To each computer sub-half parameter there is a program variable to which it is to be compared in order to assess the degradation of performance from the asymptote. The ratio of the program variable to the corresponding sub-half parameter is a dimensionless ratio, the value of which determines the extent of performance degradation that takes place. We consider here two causes of performance degradation:

1. Vectors too short :- program variable n (vector length), computer sub-half parameter $n_{\frac{1}{2}}$, and dimensionless ratio $n' = n/n_{\frac{1}{2}}$ which determines the effect of vector startup overhead.

2. Not enough arithmetic per memory reference :- program variable f (computational intensity), computer sub-half parameter $f_{\frac{1}{2}}$, and dimensionless ratio $f' = f/f_{\frac{1}{2}}$ which determines the extent of memory bottleneck.

By breaking up the reasons for performance degradation in this way and quantifying them, it is hoped to provide a better understanding of the reasons for poor computer performance, and assist manufacturers in recognising their overheads and reducing them.

3.3 The (r_∞,$n_{\frac{1}{2}}$) Parameters

We make no apology for introducing the (r_∞,$n_{\frac{1}{2}}$) parameters through their description of the vector pipeline effect in traditional vector computers. Not only was this the reason for their introduction in the first place, but MPPs are now appearing in which the individual nodes are vector processors, e.g. the Meiko CS-2, Fujitsu VPP500 and NEC SX4. To understand the performance of these highly parallel vector processors, it is necessary to describe correctly the operation of their vector pipelines. The behaviour of these pipelines and the vector instructions that use them can be characterised by the two parameters (r_∞,$n_{\frac{1}{2}}$) (see Hockney [40, 55, 41, 42, 43, 54] and the book of Willi Schönauer [67]), and the benchmark program RINF1.

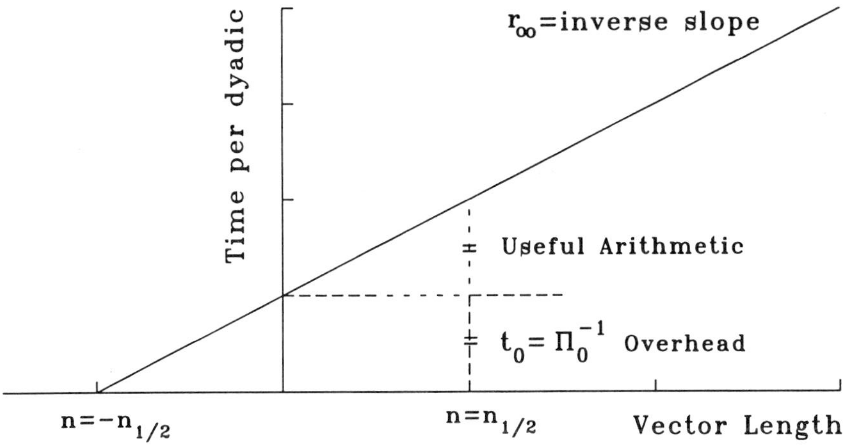

Figure 3.1: Time against vector (or Fortran DO-loop) length for a pipelined arithmetic unit, showing the geometric definition of the $(r_\infty, n_{\frac{1}{2}})$ parameters.

3.3.1 Definition of Parameters

Let us consider an arithmetic vector instruction with vector arguments of length n elements which performs a total of qn floating-point operations. Then $q = 1$ for a simple dyadic vector instruction between a pair of vectors such as $a = b \times c$ or $a = b + c$, where a, b, c are vectors each of length n elements. For more complicated vector instructions, q is the number of equivalent dyadic vector instructions, that is to say the number of dyadic vector operators (like $+$ and $\times$) in the mathematical expression using vector notation that describes the more complicated instruction. For example, $q = 2$ for the triadic instruction $a = b \times c + d$.

The definition of q is sometimes clarified by writing the vector instruction as an equivalent DO-loop which would have a loop length or number of iterations n (DO I=1,n). Then q is the number of kernel-flop in the body (or kernel) of the DO-loop, that is to say the number of flop executed in a single iteration of the loop.

Indeed it is more general to consider the $(r_\infty, n_{\frac{1}{2}})$ parameters as a linear timing model for any DO-loop (or for any linear timing relation), and there need be no connection with vector processing at all. In other words, the characterisation is equally valid whether the compiler implements the DO-loop with scalar or vector instructions, although of course the numerical value of the parameters would be quite different.

If T is the total time to execute a vector instruction for all elements (or the equivalent DO-loop for all iterations), then

$$T = q r_\infty^{-1} (n + n_{\frac{1}{2}}) \tag{3.2}$$

and we characterise the time per equivalent dyadic vector instruction (or the time per kernel-flop), $t = T/q$ as

$$\begin{aligned} t &= r_\infty^{-1}(n + n_{\frac{1}{2}}) && (3.3) \\ &= r_\infty^{-1} n + t_0 && (3.4) \end{aligned}$$

We can see from the above equations that $n_{\frac{1}{2}}$ is a way of measuring the importance of vector startup overhead ($t_0 = n_{\frac{1}{2}}/r_\infty$) in terms of quantities known to the programmer (the loop or vector length n).

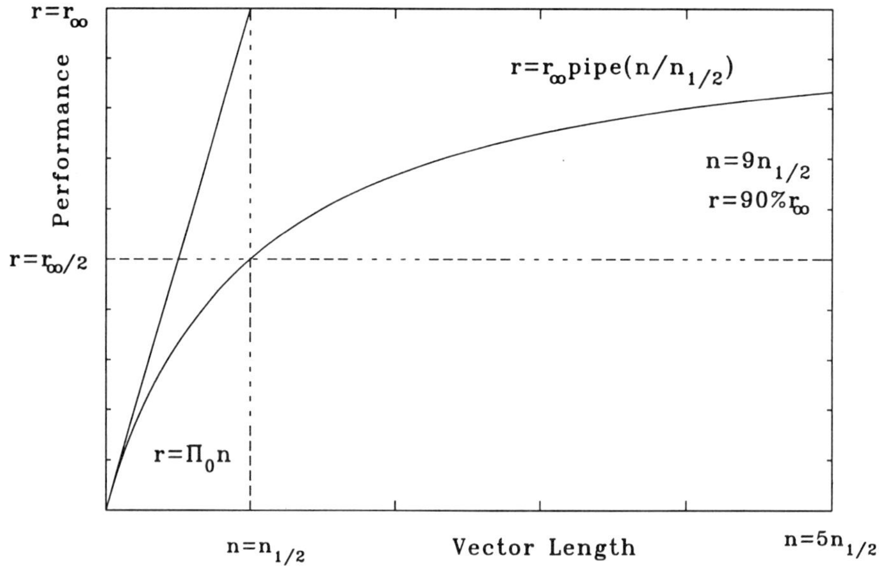

Figure 3.2: Performance against vector (or Fortran DO-loop) length for a pipelined arithmetic unit, showing the geometric definition of the $(r_\infty, n_{\frac{1}{2}})$ parameters.

Equation (3.3) and Fig. 3.1 shows that if the time per equivalent dyadic vector operation is plotted versus the vector length the result is a straight line, and that r_∞ is the inverse slope of the straight line, and $n_{\frac{1}{2}}$ is the negative of the intercept of the line with the n-axis. In the case that the measured time is not exactly linearly related to the vector length, then we take the best straight line through the measured points as approximately representing the behaviour of the vector instruction. Equation (3.3) and Fig. 3.1 also show that when $n = n_{\frac{1}{2}}$, half the time is spent on useful arithmetic (first term) and half the time is spent in startup overhead (second term).

Given Eqn.(3.3) the performance, r, is given by

$$r = \frac{qn}{T} = \frac{n}{t} = \frac{r_\infty}{(1 + n_{\frac{1}{2}}/n)} \tag{3.5}$$

Showing that r_∞ is the asymptotic performance that is approached as the vector length goes to infinity, and $n_{\frac{1}{2}}$ (called the half-performance length) is the vector length needed to achieve half that asymptotic performance. Equation (3.5) is plotted in Fig. 3.2 which shows the geometrical construction relating the parameters $(r_\infty, n_{\frac{1}{2}})$ to the performance curve. Initially for small n the slope of the performance curve is tangent to the line from the origin to the point $(n_{\frac{1}{2}}, r_\infty)$, and for large n the performance curve asymptotes to the horizontal line at $r = r_\infty$.

Although $n_{\frac{1}{2}}$ is defined in terms of half the asymptotic performance, it is important to realise that its value gives the performance for all vector lengths by use of Eqn.(3.5). The above functional form occurs so frequently in the discussion of performance that we define the pipe function as

$$\text{pipe}(x) = \frac{1}{(1 + 1/x)} \tag{3.6}$$

then

$$r = r_\infty \, \text{pipe}(n/n_{\frac{1}{2}}) \tag{3.7}$$

This can be expressed in dimensionless form as

$$r' = \text{pipe}(n') \tag{3.8}$$

where the dimensionless performance is $r' = r/r_\infty$, and the dimensionless vector length is $n' = n/n_{\frac{1}{2}}$.

A convenient consistent set of units (shown in square brackets) for the above quantities are: T [μs], t [μs/ flop], r_∞[Mflop/s], q [flop], with n and $n_{\frac{1}{2}}$ dimensionless [1].

The significance of the performance parameter $n_{\frac{1}{2}}$ is that it gives a yardstick to which the vector length can be compared, in order to decide whether one is in the short or long vector limits.

3.3.2 Long-Vector Limit

If $n \gg n_{\frac{1}{2}}$ (or the dimensionless ratio $n' \gg 1$) then we say that we are computing in the long-vector limit. In this case the second term in Eqn.(3.3) can be neglected compared with the first, and

$$t = n/r_\infty \tag{3.9}$$

$$r = n/t = r_\infty \tag{3.10}$$

Thus, as expected, the time is proportional to vector length, and the long-vector performance is a constant characterised by the parameter r_∞. This timing relation (3.9) is characteristic of serial computation (time directly proportional to work), so that algorithms optimised for serial computation will perform well on vector computers in the long-vector limit. This arises when the problem size (in the sense of vector length) is much larger than the computer (as measured by its value of $n_{\frac{1}{2}}$).

3.3.3 Short-Vector Limit

If, on the other hand, $n \ll n_{\frac{1}{2}}$ (or $n' \ll 1$), it is best to use an alternative timing expression that is algebraically equivalent to Eqn.(3.3), namely

$$t = \pi_0^{-1}(1 + n/n_{\frac{1}{2}}) \tag{3.11}$$

where $\pi_0 = r_\infty/n_{\frac{1}{2}} = t_0^{-1}$, is called the specific performance. Using these $(\pi_0, n_{\frac{1}{2}})$ parameters instead of (r_∞,$n_{\frac{1}{2}}$) the performance can be expressed in general as

$$r = n/t = \frac{\pi_0 n}{(1 + n/n_{\frac{1}{2}})} = \pi_0 n \,\text{pipe}(n_{\frac{1}{2}}/n) \tag{3.12}$$

Then in the short-vector limit the second term in the parentheses can be neglected, and

$$t = \pi_0^{-1} \tag{3.13}$$

$$r = n/t = \pi_0 n \tag{3.14}$$

Thus the time of execution is a constant independent of vector length, and the short-vector performance is proportional to vector length and characterised by π_0 (with units [Mflop/s]) rather than r_∞. This timing behaviour is characteristic of infinite parallel arrays of processors (e.g. the theoretic paracomputer often assumed during the development of parallel algorithms), so that algorithms developed for such arrays will perform well on vector computers when they are used in the short-vector limit. This situation arises when the problem is much smaller than the computer, in the sense described above.

Another useful guide to the significance of $n_{\frac{1}{2}}$ is to note from Eqn.(3.5) and Fig.3.2 that 80% of the asymptotic performance is achieved for vectors of length $4 \times n_{\frac{1}{2}}$, and that 90% of the asymptote is reached when $n = 9 \times n_{\frac{1}{2}}$. We note that this is a very slow approach to the maximum performance which is shown in Fig. 3.2. Generally speaking, $n_{\frac{1}{2}}$ values of up to about 50 are tolerable, whereas the performance of computers with larger values of $n_{\frac{1}{2}}$ is severely constrained by the need to keep vector lengths significantly longer than $n_{\frac{1}{2}}$. This requirement makes computers difficult to program efficiently, and often leads to disappointing performance, compared to the asymptotic rate advertised by the manufacturer.

3.3.4 Effect of Replication

Many parallel and vector computers are based on the replication of vector nodes or pipelines, e.g. the Meiko CS-2, the Fujitsu VPP500 and NEC SX4. The question therefore arises as to the effect of replication on the values of the performance parameters. Let a single pipeline be characterised by the parameters (r_∞,$n_{\frac{1}{2}}$) and suppose that a new computer is built with p such pipelines operating in parallel. The timing relation (3.3) will now be the time, t_m, for a vector of length $m = np$, n elements from each of p pipes.

$$t_m = r_\infty^{-1}(n + n_{\frac{1}{2}}) \tag{3.15}$$

Making the substitution $n = m/p$, we can put the timing relation in the standard form

$$t_m = r_\infty^{-1}\left(\frac{m}{p} + n_{\frac{1}{2}}\right) \tag{3.16}$$

$$t_m = (pr_\infty)^{-1}(m + pn_{\frac{1}{2}}) \tag{3.17}$$

If the p-pipeline computer has the parameters (r'_∞,$n'_{\frac{1}{2}}$) , then by definition

$$t_m = (r'_\infty)^{-1}(m + n'_{\frac{1}{2}}) \tag{3.18}$$

whence by comparison with Eqn.(3.17)

$$r'_\infty = pr_\infty \tag{3.19}$$

$$n'_{\frac{1}{2}} = pn_{\frac{1}{2}} \tag{3.20}$$

Thus we see that replication p-times multiplies both the asymptotic performance and the half-performance length by p. Taking the ratio of these last two equations, however, we find that the specific performance remains unchanged.

$$\pi'_0 = r'_\infty/n'_{\frac{1}{2}} = \pi_0 \tag{3.21}$$

Thus we find that, although the long-vector performance, r_∞, is increased p-fold, the short-vector performance, π_0, is unchanged. This is the reason for the disappointing performance of multiple pipeline computers on problems that are too small compared to the increased value of $n_{\frac{1}{2}}$.

When the above result is applied to MPPs we can conclude that the $n_{\frac{1}{2}}$ of a p-processor MPP is p times the $n_{\frac{1}{2}}$ of a single processor. With numbers of processors now approaching,

```
C  SKELETON CODE FOR RINF1 BENCHMARK
C  ---------------------------------
C
         COMMON A(100000),B(100000),C(100000)
C
C  Measure timing overhead
         CALL SECOND(T1)
         CALL SECOND(T2)
         TO=T2-T1
C
C  Main loop over vector length
         DO 20 N=1,NMAX
               DO 30 J=1,NITER
                         CALL SECOND(T1)
C LOUSY CLOCK                                       DO 30 J=1,NITER
C LOUSY CLOCK                                       CALL DUMMY(J)
C  The vector instruction
                         DO 10 I=1,N
10                       A(I)=B(I)*C(I)
C LOUSY CLOCK                                    30 CONTINUE
                         CALL SECOND(T2)
                         T=T2-T1-TO
                         TN=MIN(T,TN)
30               CONTINUE
C LOUSY CLOCK                                       TN=(T2-T1-TO)/NITER
C
C  Update least square fit
                 CALL LSTSQ(N,TN,RINF,XNHALF)
                 WRITE() N,TN,RINF,XNHALF
C
20        CONTINUE
```

Figure 3.3: Skeleton code for the RINF1 benchmark.

typically, 1000 the $n_{\frac{1}{2}}$ of an MPP is likely to be of the order 1000 to 10 000. This has unfortunate consequencies, because the value of $n_{\frac{1}{2}}$ can be identified with the speciality of the computer, with higher values corresponding with more special-purpose computers (see section 1.3.2, and reference [54] page 92-95) in the sense that there is a more limited set of problems that such computers can solve efficiently.

3.4 RINF1 Arithmetic Benchmark

The performance parameters (r_∞,$n_{\frac{1}{2}}$) should be regarded as properties of a computer which are to be measured, and of course properties of the operating system and compilers through which it is used. The RINF1 benchmark is a small program that performs this measurement for 17 different `DO-loop` kernels. A skeleton version of the program for one kernel is shown in Fig. 3.3. The subroutine `SECOND` delivers in its output parameter the elapsed wall-clock time just after the call of the subroutine. The difference in time measured by two successive calls to the subroutine is recorded in `TO`. This is the overhead of actually making the timing measurement, and it is very important to subtract this from subsequent measurements otherwise the overhead will incorrectly increase the vale of $n_{\frac{1}{2}}$. The `DO-10` loop is the kernel loop that is to be timed. It will be replaced by a vector instruction by a vectorising compiler, or it may be executed by scalar code by inserting a compiler directive. The `DO-20` loop varies the vector length, calls the timer, executes the kernel loop, and calls the timer again. The difference in time minus the overhead `TO` is recorded as the time for the vector operation. This is input to the `LSTSQ` subroutine that updates the least-squares straight-line fit, and outputs the revised values of $(r_\infty, n_{\frac{1}{2}})$ in the variables `RINF` and `XNHALF`, which are then printed.

If an accurate timer is available, preferably ticking every clock period of the computer, then the `DO-30` repeat loop can be inserted to repeat the measurement perhaps 1000 times. The minimum time found is recorded as the time for the loop. This procedure allows accurate measurements to be obtained even on busy shared memory systems, such as the Cray C-90, on which it is not usually easy to obtain sole-user access. The problem is that the arrays `A,B,C` are stored in shared memory, and access to elements may be delayed by variable amounts due to the memory banks being busy servicing the requests of other users. However if the minimum of a large number of trials is taken, there is a good chance that this time was obtained without interference from other users.

The above procedure will only work if the clock is accurate enough to measure the time for a single execution of the kernel loop to say 10% or better. Unfortunately most Unix workstations only provide clocks that tick every 10ms, and in this case the repeat loop must be placed immediately outside the `DO-10` loop in order to provide a time greater than, say, 100 ticks. This alternative `DO-30` loop in shown in comment cards starting with "`C LOUSY CLOCK`". Fortunately it will normally be possible to perform benchmark measurements on such workstations as a sole-user, and the problem of interference from other users does not arise. Most advanced optimising compilers will, however, notice that the `DO-30` loop does nothing to the numbers calculated, and will remove it. Hence it is necessary to include the call to the subroutine `DUMMY(J)` that uses the `DO-30` loop index. Since the compiler cannot know the interior of `DUMMY` (it might be separately compiled), and it could change the contents of `J`, or of `B` and `C` which are in common, neither `DUMMY` nor the `DO-30` loop can safely be removed by an optimiser.

We can see from the above that the best timing procedure to use within the RINF1 benchmark depends on the precision of the timer that should already have been determined using the TICK1 benchmark (see section 3.1). Due to the current prevalence of poor timers, the first release of the Parkbench benchmarks uses the `LOUSY CLOCK` method, but modifications of the source code can easily be made to implement the method for accurate clocks if this is desired.

3.4.1 Running RINF1

The RINF1 benchmark produces detailed output for each of 17 DO-loop kernels, followed by a summary of $(r_\infty, n_{\frac{1}{2}})$ values. Since the value of $(r_\infty, n_{\frac{1}{2}})$ depends on the operations being performed and on the size of the vectors compared to the size of the cache memory, the summary gives values for the parameter pair $(r_\infty, n_{\frac{1}{2}})$ for vector lengths that fit into the cache memory and for those that exceed the cache memory. It also records for each kernel the minimum and maximum observed Mflop/s rate for the 46 different values of loop length that are measured.

Figure 3.4 shows the detailed output for the contiguous element-by-element vector multiply kernel on IBM RS/6000 model 590 workstation, in which some of the output lines have been omitted to fit the listing on one page, and Fig. 3.5 shows the summary output for all 17 kernels from the same computer run. The detailed values of `TI` are plotted versus `NI` in Fig. 3.6 for all loop lengths, with the measured values joined by a solid line and the straight-

```
==================================================
===                                            ===
===       PARKBENCH Parallel Benchmarks        ===
===                                            ===
===                 RINF1                      ===
===                                            ===
===   Program:  R-infinity and N-half          ===
===   Version:  Standard Fortran 77            ===
===   Author:   Roger Hockney                  ===
===   Update:   November 1993; Release: 1.0    ===
===                                            ===
==================================================

      This run started Wed Sep  7 09:39:04 1994

      Run on single IBM RS/6000 Model 590
      at Old Dominion University, VA, USA (CCPO)
      Software levels are AIX 3.2.5

(1) CONTIGUOUS DYADS:  A(I)=B(I)*C(I)

ICASE=   1   NCASE=  17   NMAX=100000   NTIMES=    100000
```

SI	NI vlen	TI sec	RINF Mflop/s	N1/2 vlen	ERROR %	REPEAT	TOTAL TIME sec	R(N) Mflop/s
1	1	4.448E-08	.000	.000	.000	100000	.444E-02	22.478
2	2	9.008E-08	21.932	-.024	.000	50000	.450E-02	22.202
4	4	1.624E-07	24.980	.208	3.401	25000	.406E-02	24.629
6	6	2.527E-07	24.450	.159	1.822	16666	.421E-02	23.739
8	8	3.774E-07	21.829	-.163	2.888	12500	.471E-02	21.197
10	10	4.980E-07	20.068	-.430	3.076	10000	.498E-02	20.076
11	20	1.101E-06	17.885	-.920	2.244	5000	.550E-02	18.150
13	40	2.308E-06	17.003	-1.175	1.183	2500	.577E-02	17.329
15	60	3.493E-06	16.847	-1.238	.758	1666	.582E-02	17.174
17	80	4.720E-06	16.740	-1.302	.560	1250	.590E-02	16.948
19	100	5.928E-06	16.714	-1.325	.437	1000	.592E-02	16.868
20	200	1.196E-05	16.641	-1.428	222.200	500	.598E-02	16.717
24	600	3.593E-05	.000	.000	.000	166	.596E-02	16.697
26	800	4.788E-05	16.733	2.570	.225	125	.598E-02	16.706
28	1000	6.023E-05	16.478	-7.658	.179	100	.602E-02	16.600
29	2000	1.204E-04	16.563	-3.672	.089	50	.602E-02	16.597
31	4000	2.444E-04	16.344	-16.425	.196	25	.611E-02	16.361
33	6000	3.902E-04	15.669	-72.759	1.338	16	.624E-02	15.374
35	8000	6.708E-04	12.553	-399.009	5.538	12	.804E-02	11.925
37	10000	8.260E-04	11.914	-490.298	4.545	10	.826E-02	12.106
38	20000	1.648E-03	11.856	-504.829	2.203	5	.824E-02	12.135
40	40000	3.250E-03	12.093	-424.242	1.091	2	.650E-02	12.307
42	60000	4.915E-03	12.151	-399.903	.707	1	.491E-02	12.205
44	80000	6.532E-03	12.200	-369.560	.532	1	.653E-02	12.247
46	100000	8.162E-03	12.201	-368.505	.407	1	.816E-02	12.251

Figure 3.4: Sample detailed output from the RINF1 benchmark for the vector dyadic multiply kernel. The output has been slightly edited to fit neatly on this page.

```
          LENGTHS   RMSERR/VALUE  R-INFINITY     N-HALF        R(N)
          vlen           %          Mflop/s       vlen        Mflop/s

    (1) CONTIGUOUS DYADS:  A(I)=B(I)*C(I)
        <=    80        .560         16.740      -1.302 | Min =  11.926
        >=   600        .407         12.201    -368.505 | Max =  24.630
    (2) DYADS,  STRIDE=8:  A(I)=B(I)*C(I)
        <=     8       2.670         20.538       -.488 | Min =   1.613
        >=    60       1.479          1.577    -190.004 | Max =  24.059
    (3) CONTIGUOUS TRIADS:  A(I)=B(I)*C(I)+D(I)
        <=   800        .193         66.835       2.767 | Min =  20.078
        >=  6000        .334         20.716      47.415 | Max =  67.123
    (4) TRIADS,  STRIDE=8:  A(I)=B(I)*C(I)+D(I)
        <=   200        .328         65.435       2.057 | Min =   2.590
        >=   900        .263          2.597     -25.630 | Max =  64.974
    (5) RANDOM SCATTER/GATHER:
        <=   900        .063         18.958        .779 | Min =   9.489
        >=  7000        .668         12.418   -1354.931 | Max =  18.964
    (6) CONTIGUOUS 4-OP:  A(I)=B(I)*C(I)+D(I)*E(I)+F(I)
        <=    80        .350         44.585       -.624 | Min =  23.258
        >=   600        .319         23.313    -452.546 | Max =  55.700
    (7) INNER PRODUCT:  S=S+B(I)*C(I)
        <=   700        .253         66.474        .529 | Min =  34.962
        >=  5000        .986         34.204   -2090.563 | Max =  67.506
    (8) FIRST ORDER RECURRENCE:  A(I)=B(I)*A(I-1)+D(I)
        <=    80        .746         33.781      -1.911 | Min =  23.197
        >=   600        .734         23.911    -454.478 | Max = 133.570
    (9) CHARGE ASSIGNMENT:  A(J(I))=A(J(I))+S
        <=    80        .202         11.392       -.640 | Min =  10.796
        >=   600        .296         10.815    -278.668 | Max =  24.193
    (10) TRANSPOSITION:  B(I,J)=A(J,I)
        <=    60        .230         32.877        .912 | Min =  16.660
        >=   400        .000         26.938      -1.743 | Max =  32.430
    (11) MATRIX MULT BY INNER PRODUCT
        <=    80        .585         65.653       2.301 | Min =  13.276
        >=   600        .000         51.412      -4.345 | Max =  64.166
    (12) MATRIX MULT BY MIDDLE PRODUCT
        <=    80        .481         66.191       1.944 | Min =  12.105
        >=   600        .000         52.649      -4.183 | Max =  64.473
    (13) MATRIX MULT BY OUTER PRODUCT
        <=  1600        .157         31.859       7.024 | Min =   7.370
        >=   200        .000         12.834     179.173 | Max =  32.181
    (14) DYADS,  STRIDE=128:  A(I)=B(I)*C(I)
        <=     0        .000           .000        .000 | Min =    .873
        >=     6       4.035           .828     -35.583 | Max =  23.339
    (15) DYADS,  STRIDE=1024:  A(I)=B(I)*C(I)
        <=     4       2.496         24.984        .223 | Min =    .885
        >=    20        .033           .885       -.009 | Max =  24.311
    (16) CONTIGUOUS DAXPY:  A(I)=S*B(I)+C(I)
        <=    80        .548         33.508      -1.271 | Min =  23.192
        >=   600        .621         24.181    -428.602 | Max =  48.118
    (17) INDIRECT DAXPY:  A(J(I))=S*B(K(I))+C(L(I))
        <=   600        .077         26.552        .436 | Min =   5.406
        >=  4000        .086          5.427      -7.197 | Max =  26.572
```

Figure 3.5: Summary output from the RINF1 benchmark for the 17 kernels. The output has been slightly edited to fit neatly on this page

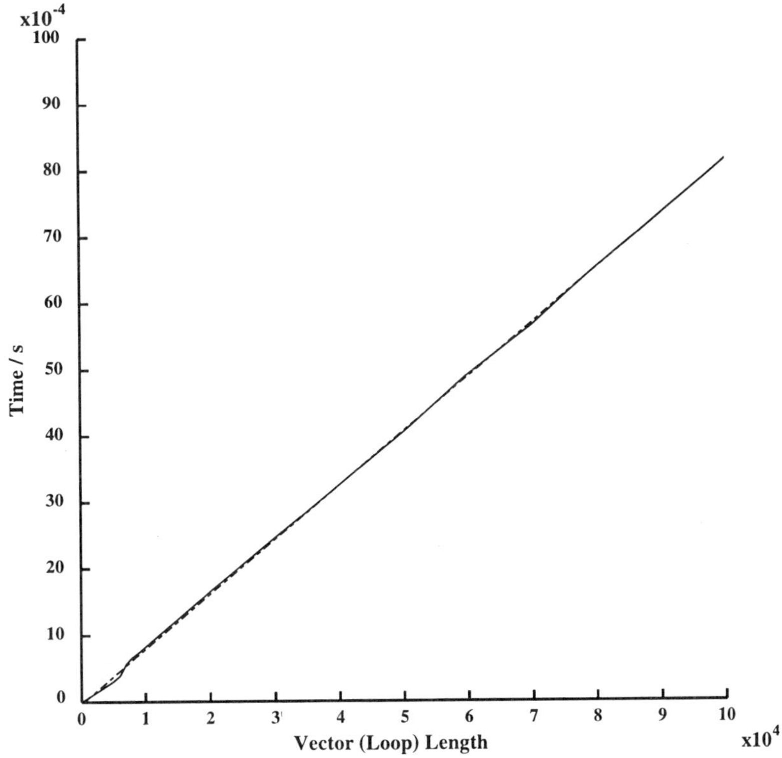

Figure 3.6: Time against vector length for the RINF1 contiguous dyadic multiply kernel on the IBM RS/6000, showing all vector lengths

line fit corresponding to the last values of (r_∞,$n_{\frac{1}{2}}$) drawn as a dotted line (r_∞=12.1 Mflop/s, and$n_{\frac{1}{2}}$=-368). The quality of the fit is good for long loop lengths greater than n = 10 000 because the two lines can scarcely be distinguished, but below n = 10 000 it is clear that something different is happening and that a single straight-line fit is inadequate for all loop lengths.

Cache effects

The region up to n = 10 000 is plotted separately in Fig. 3.7, and shows a good straight-line fit for $n < 4000$ with r_∞ = 16.7 and $n_{\frac{1}{2}}$ essentially zero, followed by a transitional region up to about n = 20 000 when the fit to the long loop-length values becomes good. We associate this phenomenon with the increase in the number of cache misses—and therefore an increase in computer time—that arises as the data required by the DO-loop begins to exceed the capacity of the data cache. Because the DO-loop is repeated many (in fact `NITER`) times, if all the data fits into the cache then, except for the first execution of the loop, the required data will be found in the cache, and we have an *in-cache* measurement. On the other hand if all the data does not fit in the cache, depending on the particular caching algorithm used, different numbers of cache misses will occur, and the time will be increased over that which would arise from an extrapolation of the in-cache behaviour to longer loops. Therefore we call the straight-line fit to these longer loop lengths an *out-of-cache* measurement. These two straight-line fits are shown as the dotted lines in Fig. 3.7, and it is clear that two straight lines

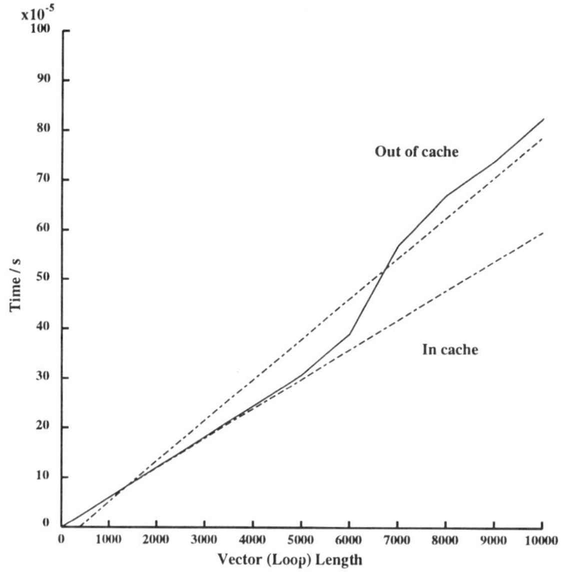

Figure 3.7: Time against vector length for the RINF1 contiguous dyadic multiply kernel on the IBM RS/6000, showing transition region between in-cache and out-of-cache situation

with a step transition at $n = 6000$ fit the measured data for the IBM RS/6000-590 very well. This behaviour is typical of cache-based RISC processors that usually show a negative value for the out-of-cache $n_{\frac{1}{2}}$, whereas traditional vector computers like the Cray C-90 are fitted quite well with a single straight line with positive $n_{\frac{1}{2}}$. For a more comprehensive discussion of cache effects, see Getov [31], and two papers by Schönauer and Häfner [68, 69].

Operating Instructions

Detailed instructions for running the Parkbench benchmarks will vary slightly from release to release as the benchmark suite is improved and made easier to use, so we give here only the general procedure. Users should consult the '`ReadMe`' files first at the top-level directory called `PARKBENCH`, and then in the subdirectory `LOWLEV`. At the top level instructions are given for setting up the PVM environment variables and 'making' all the executables. At the lower level instructions are given for preparing any input data files, for saving the result files and any special instructions for a particular benchmark. The information given below is extracted from the '`ReadMe`' files of release 1.0 of the Parkbench benchmarks dated 20 June 1995.

After setting up any necessary environment variables, the executables can generally be prepared by going to directory `PARKBENCH/Makefile` and typing '`make LOWLEV`'. This makes all the lowlevel executables. A particular executable can be made by typing '`make`' followed by the benchmark name, in this case '`make rinf1`'.

Then to run the benchmark, type the benchmark name: `rinf1`

Output from the benchmark is written to file made from the benchmark name with a '`.res`' extension, in this case '`rinf1.res`'. Copy this to another file to save it, as it will be overwritten by the next execution of the benchmark.

This benchmark assumes by default that the maximum vector length is 100 000. Change the parameters `NNMAX` if this is not suitable. It is also advisable to check the number of iterations and to adjust this if necessary in accordance with the clock tick.

```
NITER = 1000   if tick is 1.0E-5 s
NITER = 100000 if tick is 1.0E-3 s
```

All parameters are to be found in the include file '`rinf1.inc`'.

If `NITER=10000` RINF1 will take about 2 minutes to run on a typical workstation. For accurate results with `NITER=100000` allow 15 to 20 minutes.

Interpretation of Results

Low-level benchmarks like RINF1 are trying to represent, for each kernel, some 50 measurements (the time for each of 50 vector lengths) by two performance parameters $(r_\infty, n_{\frac{1}{2}})$. The times to be measured are also very short, and if the repeat number `NITER` is not large enough for the timer being used, nonsense values for the time of execution will give nonsense values for the parameters. It really is a case of garbage-in gives garbage-out. In a shared-memory computer, contention for the same memory banks from other users can also give a large scatter to the measured times and give unsatisfactory results. Compared with an application benchmark that only requires the measurement of a single long time interval comparable to a second or minute, for perhaps only three input data sets, without any effort to fit the time to a model, the interpretation of data from low-level benchmarks is incomparably more difficult. Good results are not to be expected from such benchmarks unless they are carried out with care and interpreting with good sense.

The summary table of results at the end of the printed output (see e.g. Fig. 3.5) is an attempt to pick automatically from the mass of measurements the best value of the parameter pair $(r_\infty, n_{\frac{1}{2}})$ for in-cache values (reported first) and out-of-cache values (reported second). The summary line states the vector lengths that have been used to obtain these values. If the summary look silly, and perhaps in any case, one should also examine the detailed output (see e.g. Fig. 3.4), because the automatic selection cannot be expected always to work satisfactorily. These are our recommendations for interpreting the detailed results. For each of the 17 kernels (DO loops):

1. Examine the `TOTAL TIME` column of the output and ensure that this is at least 100 times the measured tick of your timer as measured by the TICK1 benchmark. If not increase `NITER` by a factor 10. The run will now take longer, but the timing results should show much less scatter.

2. Examine the values in the time column `TI`, this is the time per dyadic vector operation as a function of the vector length in the column headed `NI`. If `TI` is not a monotonically increasing function of `NI` that is roughly linear, then the $(r_\infty, n_{\frac{1}{2}})$ parameters are not appropriate, and this benchmark will not make sense. Therefore plot `TI` against `NI` and see what it looks like. If there is a lot of scatter, then increase `NITER` and rerun. If it is reasonably smooth but not at all linear, or representable as a small set of straight lines, do something else. If it is approximately linear then the columns headed `RINF` and `N1/2` should have stable values that do not change much as the vector length increases. This is what one is looking for, and such stable values are the ones to be reported.

3. The column headed `PCT ERROR` gives the root mean square deviation of the line from the measured points with the parameters derived from the data points, expressed as a percentage of the last value of `TI`. Values up to a few percent indicate that the straight-line fit is good and that the $(r_\infty, n_{\frac{1}{2}})$ values are reliable. Values greater than say 20% indicate that the approximation is poor and the parameters should be used with caution, if at

all. Bear in mind also that values of $n_{\frac{1}{2}}$ are added to n(=`NI`) in Eqn.(3.3), and divided by n in Eqn.(3.5), thus large values and variations in $n_{\frac{1}{2}}$ may in fact be insignificant and unimportant when the value of n itself is large. They do not necessarily indicate an unsatisfactory result.

4. It is important to understand the meaning of the values in the columns `RINF` and `N1/2`. The vector lengths are run through in the order printed and as a new time of execution is obtained for the next vector length, updated values of `RINF` and `N1/2` are computed. That means that the values printed on one line are the best least-squares fit of a straight line to all data computed up to this time (i.e. all `NI`, `TI` pairs appearing on this and all previous lines, but not of course from any later lines).

   The first line (`SI=1`) provides only one point and does not define a straight line, so `RINF=N1/2=0` is printed, meaning not enough information to compute values. By `SI=2` there are two points and a straight line is defined together with the values of `RINF` and `N1/2`. The fit is exact and the `ERROR` column records correctly zero. As each new point is computed, `RINF` and `N1/2` are updated, with the best least-squares straight line. For small vector lengths, and perhaps an inaccurate timer, values of `RINF` and `N1/2` may wave around and even become negative. This does not matter provided the values stabilise for longer vector lengths. It probably means that `NITER` was taken too small. Apart from the effects of cache, discussed next, the best values of `RINF` and `N1/2` should be the last ones recorded for the longest vector, because this straight line uses all the previous data values.

5. The presence of a data cache complicates the picture considerably by increasing the execution times significantly once the vector length exceeds the cache or paging size, when references to off-chip memory are required. This shows up by driving the value of `N1/2` negative as shown in Fig. 3.7. This is correct and only means that the best straight line intercepts the positive x-axis. In the sample results shown in Fig. 3.4, `RINF` and `N1/2` have stabilised before this point, and these values are the in-cache measurement.

   The automatic selection procedure tries to pick these values and prints them in the summary table. This trip point where `N1/2` goes negative is marked in the detailed output by '`PCT ERROR`' being set to 222.2. The selected value is taken three measurements before this point. The least-squares fit is then reset, and a separate best straight line is obtained for longer vectors exceeding the cache size. This provides a second pair of (r_∞,$n_{\frac{1}{2}}$) values for vectors longer than the value stated in the summary table. This value is 4 measurement points past the trip point, in order to avoid using points in the transition region.

It must be obvious from the above that sensible results will only be obtained from RINF1 if the benchmark is run sensibly (using in particular a good clock and a sensible value for `NITER`), and the results are interpreted with care and understanding. It is easy to misuse this benchmark and produce rubbish results. It is therefore also easy for anyone to "rubbish" the benchmark if they should wish, for any reason, to do so; however it delivers good understanding of the behaviour of the basic arithmetic hardware (and the software through which it is used) when it is used properly.

Negative values of r_∞ and $n_{\frac{1}{2}}$

It is often supposed that negative values for r_∞ and $n_{\frac{1}{2}}$ are meaningless and therefore bring the benchmark into disrepute. This shows a misunderstanding of the parameters: r_∞ and

$n_{\frac{1}{2}}$ should be thought of as two parameters that determine, respectively, the inverse slope and the negative intercept on the x-axis, of a straight line (see Fig. 3.1). They are used in Eqn.(3.3) and (3.5) to determine the performance, r, or time, t, as a function of vector length, n. Whereas neither r, t nor n can by their very nature be negative, there is no reason why in certain circumstances r_∞ and $n_{\frac{1}{2}}$ cannot be negative. Such negative values can appear for small values of n with an inaccurate timer, and should generally be ignored, provided later values stabilise. As explained in section 3.4.1, negative values of $n_{\frac{1}{2}}$ are quite usual and correct for out-of-cache measurements. Negative r_∞ would imply that larger problems execute in less time, and this would not be expected, but there may be such cases. In fact the benchmark traps negative r_∞ with negative $n_{\frac{1}{2}}$ as indicating poor input data, rejects such data and restarts the least squares fit. This action is signalled in the detailed output by the value of `PCT ERROR` being 111.1. The only statement that we can say with certainty is that r, t and n computed from Eqn.(3.3) and (3.5) cannot be negative. (See also Getov [31]).

3.5 COMMS Communication Benchmarks

The $(r_\infty, n_{\frac{1}{2}})$ characterisation can be used for any process that has a linear timing relation with respect to some variable, and need not necessarily have anything to do with vector processing or arithmetic. A good non-arithmetic example is the time for disk access as a function of the length of the data transferred, which has a seek time followed by a time per element. A similar example is the time for communication between nodes on a distributed memory computer as a function of message length, and we now use the parameters for just this case. However, before communication performance can be characterised it must be measured, and that is the purpose of the COMMS benchmarks.

3.5.1 COMMS1: Pingpong or Echo Benchmark

In the COMMS1, *Pingpong* [45, 46] or Echo [25] benchmark we take a master and a slave node, and do all the timing on the master node with a single clock. This avoids the problem of clock synchronisation that would arise if the start of the message was timed on one processor and its receipt was timed with a different clock on another processor. A message of a given length is sent from the master node to the slave node, and when it has been received into a Fortran user array on the slave, it is immediately returned to the master. The timing ends when the complete array of data is available for use in the master. Half the time for this message pingpong is recorded as the time to send a message of the given length. The length is varied and a least-squares straight-line fit to the time against message length data gives the values of $(r^c_\infty, n^c_{\frac{1}{2}})$ as in the RINF1 benchmark. We use a superscript 'c' on these and other variables to show that they are the values obtained to characterise communication. Then we have the time to send a message as

$$t^c = (n^c + n^c_{\frac{1}{2}})/r^c_\infty = t^c_0 + n^c/r^c_\infty \tag{3.22}$$

or alternatively

$$t^c = (1 + n^c/n^c_{\frac{1}{2}})/\pi^c_0 \tag{3.23}$$

and the realised communication rate is given by

$$r^c = \frac{n^c}{t^c} = \frac{r^c_\infty}{1 + n^c_{\frac{1}{2}}/n^c} = r^c_\infty \,\mathrm{pipe}(n^c/n^c_{\frac{1}{2}}) \tag{3.24}$$

or alternatively, for small n^c it is better to use the algebraically equivalent expression

$$r^c = \frac{\pi_0^c n^c}{1 + n^c/n_{\frac{1}{2}}^c} = \pi_0^c n^c \, \mathrm{pipe}(n_{\frac{1}{2}}^c/n^c) \tag{3.25}$$

where the startup time (also called the message latency) is

$$t_0^c = 1/\pi_0^c = n_{\frac{1}{2}}^c/r_\infty^c \tag{3.26}$$

In the above equations, r_∞^c is the *asymptotic bandwidth or stream rate* of communication which is approached as the message length tends to infinity, and $n_{\frac{1}{2}}^c$ is the message length required to achieve half this asymptotic rate. Hence $n_{\frac{1}{2}}^c$ is called the *half-performance message length*. Manufacturers of distributed-memory message-passing computers normally publish the r_∞^c of the communication, but will rarely quote any values for the message startup time, or equivalently the value of $n_{\frac{1}{2}}^c$. Since the realised bandwidth for messages less than $n_{\frac{1}{2}}^c$ in length is dominated by the startup time rather than the stream rate, it is particularly important that the value of $n_{\frac{1}{2}}^c$ be known. It is to be hoped that manufacturers will be encouraged to measure and quote both the parameters $(r_\infty^c, n_{\frac{1}{2}}^c)$, and the derived parameter π_0^c, in order to enable a proper assessment of their communication hardware and software.

In analogy with $n_{\frac{1}{2}}$ for arithmetic (section 3.4), the importance of the parameter $n_{\frac{1}{2}}^c$ is that it provides a yardstick with which to measure message-length, and thereby enables one to distinguish the two regimes of short and long messages. For long messages $(n^c \gg n_{\frac{1}{2}}^c)$, the denominator in equation (3.24) is approximately unity and the communication rate is approximately constant at its asymptotic rate, r_∞^c

$$r^c \approx r_\infty^c \tag{3.27}$$

For short messages $(n^c \ll n_{\frac{1}{2}}^c)$, the denominator in Eqn.(3.25) is approximately unity, so that

$$r^c \approx \pi_0^c n^c = n^c/t_0 \tag{3.28}$$

In sharp contrast to the approximately constant rate in the long-message limit, the communication rate in the short message limit is seen to be approximately proportional to the message length, and the constant of proportionality, π_0^c, is known as the *specific bandwidth.* Thus, in general, we may say that r_∞^c characterises the long-message performance and π_0^c the short-message performance.

Because of the finite (and often large) value of t_0^c, the above is a *two-parameter* description of communication performance. It is therefore incorrect, and sometimes positively misleading, to quote only one of the parameters (e.g. just r_∞^c, as is often done) to describe the performance. The most useful pairs of parameters are $(r_\infty^c, n_{\frac{1}{2}}^c)$, $(\pi_0^c, n_{\frac{1}{2}}^c)$ and (t_0^c, r_∞^c), depending on whether one is concerned with long messages, short messages or a direct comparison with the characteristic times of the hardware.

Some authors, notably Dunigan [25, 26], have used a similar two-parameter description and express the time to transmit a message as

$$t = \alpha + \beta n \tag{3.29}$$

where α is the startup time, and β is the byte-transfer time. Comparing Eqns.(3.22) and (3.29) shows that the two models are exactly the same, and that there is the following correspondence between the two notations:

$$\alpha = t_0^c = 1/\pi_0^c = n_{\frac{1}{2}}^c/r_\infty^c; \qquad \beta = 1/r_\infty^c \tag{3.30}$$

and

$$r_\infty^c = 1/\beta; \qquad n_{\frac{1}{2}}^c = \alpha/\beta \tag{3.31}$$

In the case that there are different modes of transmission for messages shorter or longer than a certain length, the benchmark can read in this breakpoint and perform a separate least-squares fit for the two regions. An example is the Intel iPSC/860 which has a different message protocol for messages shorter than and longer than 100 byte.

3.5.2 COMMS2 Benchmark

The COMMS2 benchmark is very similar to COMMS1 but measures the message exchange properties of a computer network. A pair of nodes send a message of varying length to each other and then wait to receive the message from the other of the pair. One quarter of the time for this exchange is recorded as the time to send a message, because four messages are sent during the exchange. In order to avoid memory conflicts between data arriving and data being sent, different arrays are used as arguments to the send and receive operations. In this data exchange, advantage can be taken of bidirectional links if they exist, and in this case a greater bandwidth can be obtained than is possible with COMMS1. The operating instructions and output file format are the same as for COMMS1.

Data exchange is an important operation in many algorithms for the solution of partial differential equations by domain decomposition. In these, a large region is decomposed into many smaller regions, each of which is assigned to a different processor. During the solution boundary data must be exchanged across the surface of each region to the processor responsible for the neighbouring region. Similarly in particle-in-cell simulation codes (see e.g. section2.7), the coordinates of particles which cross from one region to another must be exchanged at each timestep. For these reasons the message-exchange performance that is measured by COMMS2 is of considerable importance.

3.5.3 Running the COMMS1 and COMMS2 Benchmarks

The COMMS1 benchmark has been deliberately kept simple by restricting the test to asynchronous communication. This is the most favourable case and gives a lower bound on the time for the communication of a message. Asynchronous, here, means that a send returns to the calling program when the user data array being sent may be safely reused. This, however, may be before the message has been received by the receiving node. The receiving node program blocks (i.e. stops) at the corresponding receive instruction until the data is available for use by the user's program.

Operating Instructions

To compile and link the code with the appropriate libraries for PVM, enter the directory `pvm3` and type: `make`

On some systems it may be necessary to allocate the appropriate resources before running the benchmark, eg. on the iPSC/860 to reserve a cube of 2 processors, type: `getcube -t2`

The message length of each test is defined by a file called '`msglen.def`'. If you wish to obtain a benchmark result for comparison with results from other machines, you should use the standard version of `msglen.def` provided with this release. Alternatively, if you wish to investigate the detailed variation of communication speed with message length for your particular machine, you can edit the file before running the benchmark. The format is one integer value per line, each defining the message length of a test case. The values should be in ascending order. You can specify any number of values, up to a compile time limit specified by the parameter **MAXTST**, which is defined in the file '`comms1.inc`'.

Further input data that controls the running of the benchmark is contained in the file '`comms1.dat`' which "answers" various questions posed by the running program. This file should be edited according to the needs of the benchmarker.

The first line of this file gives the number processors (or nodes), **NNODE**, to be allocated. The default is 2, but you may allocate any number up to a maximum defined by the compile-time parameter **MAXNOD**, declared in the file '`comms1.inc`'. If you choose more than two processors, you can specify which slave node, **NSLAVE**, you want the master (node 0) to communicate with. This can be any number between 1 and **NNODE-1**. This option can be used to study the time variation with separation within the network. Many message-passing computers have different timing for short and long messages, and the next line in the '`.dat`' file gives the number of bytes, **NSBYTE**, in the longest short message, or zero if there is no difference between short and long messages. If you specify a non-zero value, the program will automatically add test cases for the longest short message and the shortest long message, if they are not already defined in '`msglen.def`'. The next data line specifies whether or not timings for zero length messages are to be used in computing least squares fits to the data. A value of zero includes the zero length data, and a value of one excludes it. This is useful since such timings can be anomalous. Finally the last line of the data file specifies **MTIME**, the approximate execution time it is desired that the measurement for each message length will take. The actual number of times a message is pingponged for each case is calculated to give approximately that execution time. This means that, for any particular system, you can ensure each test is run for long enough to average out disturbances caused by spurious operating system effects. It also means that you have direct control over the total time the benchmark will take to run.

When the above data has been read, the program proceeds to make estimates of the loop overhead and communication parameters. It uses these to calculate the number of pingpongs needed for each test case, to obtain the requested execution time per test. It should be noted that the loop overhead is re-measured for each test, and that the measurement takes approximately the same time as the pingpong part of the test, so the total elapsed time for each test case is actually about twice the specified execution time.

To run the benchmark executable, type: `comms1`

This will automatically load both host and node programs. Once the timing parameters have been estimated, the benchmark test cases are executed. To enable their progress to be followed, a line is written to the standard output, showing the test number and message length, when each test starts. When it finishes, the measurement of the time taken to send one message is written out, together with the number of iterations the test used. A permanent copy of the full benchmark results is written to a file called '`comms1.res`'. If the run is successful and a permanent record is required, this file should be copied to another file before the next run overwrites it.

Sample Results

Figure 3.8 shows the printer output file for COMMS1 when run under PVM on the Meiko CS-2. Column-2 gives the message transmission time measured for 46 different values of message length which are specified in an input file (in the figure some output lines have been omitted to allow the file to fit on one page). Columns-4 and -5 give the asymptotic bandwidth, r_∞^c in units of [MB/s] and $n_{\frac{1}{2}}^c$ in units of [B] respectively, and column-6 gives the root mean square error of the straight-line fit as a percentage of the currently measured time, in exactly the same way as in the RINF1 benchmark. The value is less than a percent in this column which shows that the fit is very good in this case. The behaviour for short messages can best be compared

```
=================================================
===  GENESIS Distributed Memory Benchmarks    ===
===                                           ===
===               COMMS1                      ===
===                                           ===
===    Program:   Message Pingpong            ===
===    Version:   PVM 3.1 Fortran 77          ===
===    Authors:   Roger W. Hockney            ===
===               Ian Glendinning             ===
===               Ade Miller                  ===
===    Updates:   Jun 1994 - Release 3.0      ===
=================================================
Run on Meiko CS-2 at the University of Southampton
Compiler version: SC2.0.1 20 Apr 1993 Sun FORTRAN
2.0.1 patch 100963-03, b/end SC2.0.1 03 Sep 1992
Operating system: SunOS 5.1 MEIKO FCS

The measurement time requested for each test case was  1.0 seconds
No distinction was made between long and short messages
Zero length messages were used in least squares fitting

   Case  LENGTH(B)   TIME(sec)   RINF(B/s)    N1/2(B)     %error fit
     1        0      1.983E-04   0.000E+00   0.000E+00   0.000E+00
     2        1      2.001E-04   5.329E+05   1.056E+02   0.000E+00
     3        2      2.006E-04   8.383E+05   1.664E+02   1.606E-01
     4        4      2.007E-04   1.886E+06   3.752E+02   3.000E-01
     5        6      2.014E-04   2.419E+06   4.817E+02   2.845E-01
     6        8      2.023E-04   2.487E+06   4.953E+02   2.590E-01
     7       10      2.010E-04   3.910E+06   7.799E+02   3.761E-01
     8       20      2.004E-04   1.569E+07   3.140E+03   5.083E-01
    10       40      2.025E-04   1.878E+07   3.761E+03   4.551E-01
    12       60      2.043E-04   1.512E+07   3.027E+03   4.228E-01
    14       80      2.048E-04   1.467E+07   2.936E+03   4.292E-01
    16      100      2.070E-04   1.390E+07   2.780E+03   4.189E-01
    26      200      2.206E-04   9.203E+06   1.832E+03   8.176E-01
    37      400      2.420E-04   9.149E+06   1.821E+03   6.503E-01
    38      600      2.567E-04   9.652E+06   1.927E+03   7.239E-01
    39      800      2.796E-04   9.822E+06   1.963E+03   6.684E-01
    40     1000      2.973E-04   1.003E+07   2.006E+03   6.455E-01
    41     2000      3.895E-04   1.041E+07   2.089E+03   5.415E-01
    42     5000      6.940E-04   1.017E+07   2.038E+03   3.233E-01
    43    10000      1.218E-03   9.897E+06   1.975E+03   2.734E-01
    44    20000      2.348E-03   9.432E+06   1.862E+03   4.238E-01
    45    30000      3.450E-03   9.295E+06   1.825E+03   3.417E-01
    46    40000      4.598E-03   9.180E+06   1.792E+03   3.325E-01
                               ----------------
                               Result Summmary
                               ---------------

  rinf = 9.180 MByte/s,  nhalf = 1791.949 Byte,  startup = 195.200 us
```

Figure 3.8: Sample output for the COMMS1 benchmark for PVM on the Meiko CS-2

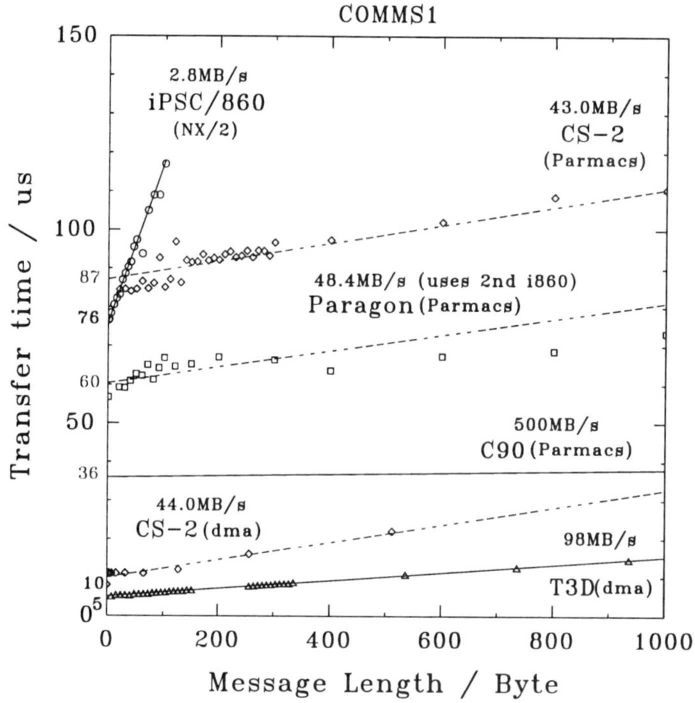

Figure 3.9: Time versus message length for a variety of computers, obtained by plotting the output from COMMS1

by plotting 'time' versus 'message length' on a linear-linear graph, since this allows one to judge the quality of the straight-line approximation. One such graph, comparing a variety of different computer communication systems and software, is shown in Fig. 3.9. Alternatively the results can be expressed in terms of the measured bandwidth as a function of the message length. The straight-line fit of t versus n corresponds to a pipeline variation of bandwidth with length (see Eqn.(3.24)). In this case a log/log graph is most appropriate because the shape of the pipe function is invariant with respect to its position on such a graph (see section 5.4.1). An example is shown in Fig. 3.10. Both types of graph can now be produced from data in the Parkbench database by using the Graphical Benchmark Information Service (GBIS) which is described in section 5.4.

3.5.4 Total Saturation Bandwidth: COMMS3

To complement the above communication benchmarks, there is a need for a benchmark to measure the total saturation bandwidth of the complete communication system, and to see how this scales with the number of processors. This benchmark attempts to measure the saturation bandwidth by swamping the communication system with messages. A natural generalisation of the COMMS2 benchmark is made as follows, and called the COMMS3 benchmark: Each processor of a p-processor system sends a message of length n to the other $(p-1)$ processors. Each processor then waits to receive the $(p-1)$ messages directed at it. The timing of this generalised *exchange* ends when all messages have been successfully received by all processors; although the process will be repeated many times to obtain an accurate measurement, and the overall time will be divided by the number of repeats. The time for the generalised exchange is the time to send $p(p-1)$ messages of length n and can be analysed in the same way as

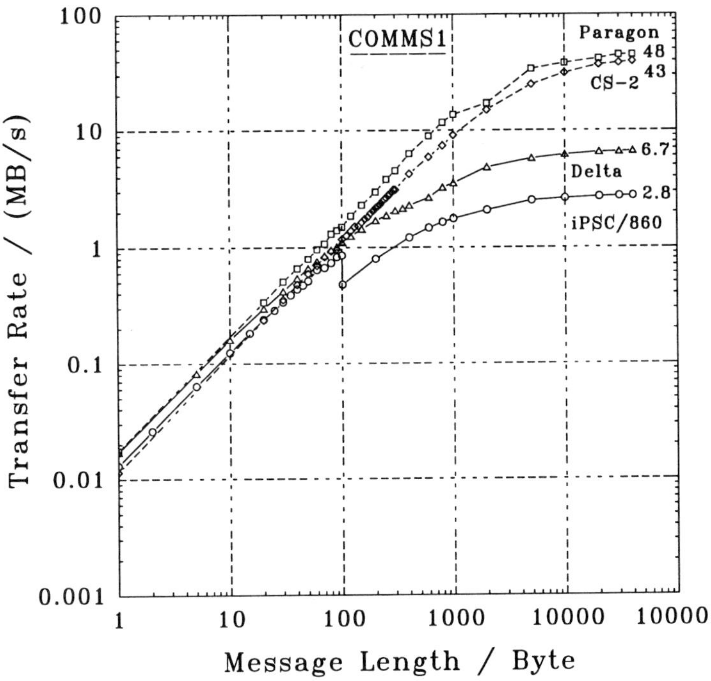

Figure 3.10: Communication bandwidth versus message length for a variety of computers, obtained by plotting the output from COMMS1. The data is the same as for Fig.3.9

COMMS1 and COMMS2 into values of $(r_\infty^c, n_{\frac{1}{2}}^c)$. The value obtained for r_∞^c is the required total saturation bandwidth, and we are interested in how this scales up as the number of processors p increases and with it the number of available links in the system. The program records the maximum observed bandwidth, and the corresponding bandwidth per processor.

Running COMMS3

First the data file controlling COMMS3 should be edited. This is '`comms3.dat`' and contains two lines. The first contains the value of **`NNODE`** which is the number of processors taking part in the test. The second line contains **`NEND`** which selects the range of n values to be used. The benchmarker is entitled to vary **`NEND`** in order to obtain the highest measured bandwidth for each value of **`NNODE`**. Then he should plot the curve of 'Maximum observed bandwidth' against 'number of processors'.

Next the parameter **`NITER`** in '`comm3.inc`' should be adjusted if necessary. This is the number of repeats of the test within the timing loop, may be increased from 1 or 10 (for short test to confirm execution) to 1000 or 10 000 (for time measurements), depending on the precision of your clock (see TICK1 benchmark, section 3.1.1).

To compile and link the code with the appropriate libraries for PVM, enter the directory `pvm3` and type: `make`

On some systems it may be necessary to allocate the appropriate resources before running the benchmark, eg. on the iPSC/860 to reserve a cube of 2 processors, type: `getcube -t2`

To run the benchmark executable, type: `comms3`

This will automatically load both host and node programs, and read the values of **`NNODE`**

```
=================================================
===  GENESIS / PARKBENCH Distributed Memory   ===
===                Benchmarks                 ===
===                                           ===
===                  COMMS3                   ===
===                                           ===
===     Program:   Saturation Bandwidth       ===
===     Version:   PVM +  Fortran 77          ===
===     Author:    Roger W. Hockney           ===
===     Update:    November 1993;             ===
=================================================
Run on Meiko CS-2 at the University of Southampton
Compiler version: SC2.0.1 20 Apr 1993 Sun FORTRAN
2.0.1 patch 100963-03, b/end SC2.0.1 03 Sep 1992
Operating system: SunOS 5.1 MEIKO FCS

      Case  LENGTH(B)   TIME(sec)   RAV(B/s)
        1       200     1.122E-04   1.783E+06
        2       400     1.036E-04   3.860E+06
        3       600     1.268E-04   4.732E+06
        4       800     1.493E-04   5.357E+06
        5      1000     1.705E-04   5.865E+06
        6      2000     2.823E-04   7.085E+06
        7      5000     6.838E-04   7.312E+06
        8     10000     1.248E-03   8.010E+06
        9     20000     2.494E-03   8.018E+06
       10     50000     6.120E-03   8.170E+06
                       --------------
                       Result Summary
                       --------------
Number of Processors(nodes) in test =   4
Maximum Observed Total Bandwidth =  8.170E+06 B/s
Maximum Bandwidth per Processor  =  2.042E+06 B/s
```

Figure 3.11: Output for the COMMS3 benchmark for PVM on the Meiko CS-2

and `NEND` from '`comms3.dat`'. The progress of the benchmark execution can be monitored via the standard output, whilst a permanent copy of the benchmark is written to a file called '`comms3.res`', which normally appears in your top directory after the run is complete. If the run is successful and a permanent record is required, the file '`comms3.res`' should be copied to another file before the next run overwrites it.

A sample output file for COMMS3 is shown in Fig. 3.11 for PVM on the Meiko CS-2. It is self-explanatory. Normally runs would be performed for a series of values for the number of processors, so that the variation of saturation bandwidth with the number of processors can be determined.

3.6 POLY or Balance Benchmarks

It is now relatively easy and cheap to make fast pipelined arithmetic VLSI chips (e.g. the Intel i860 and IBM RS/6000 chip set) but much more difficult and expensive to provide a high bandwidth for the transfer of data between the external memory and the arithmetic chip. Since a pipelined arithmetic unit requires two input arguments and one output result to be transferred every clock period, the communication bandwidth (measured in millions of words per second, Mw/s) needs to be three times faster than the arithmetic rate (measured in millions of floating-point operations per sec, Mflop/s), in order that the memory access be in balance with the arithmetic unit. Such a high memory bandwidth is, in fact, provided by the top-of-the-range vector supercomputers, e.g. the Cray C-90 and NEC SX2, and this is one reason why these machines are expensive. For vector computers lower in the range, and for parallel designs based on multiple microprocessors (e.g i860 or RS/6000 chips) this level of memory bandwidth is not normally provided, and there will exist a memory bottleneck. In this case, the performance of the system is likely to be determined more by the available memory bandwidth than by the arithmetic capability of the microprocessors.

If there is a memory access bottleneck, then the key program variable to consider is the *computational intensity*, f, which is defined as the number of arithmetic operations performed per memory transfer [54]. Put another way, it is how intensely one computes with data once it has been received in the registers (or cache) within the arithmetic unit. If the computational intensity is high, then the time spent on data transfer becomes small compared to the time spent on arithmetic, and the memory bottleneck is not seen. On the other hand, if the computational intensity is low then the time spent on data transfers dominates the calculation, and the performance of the computer is much less than that advertised on the basis of the arithmetic rate. The question is: To what computer performance parameter do we compare the computational intensity (which is a program variable), in order to assess quantitatively the meaning of high and low in the above statements?

To answer this question we consider a main memory connected to an arithmetic unit by a memory access pipeline characterised by the parameters $(r_\infty^m, n_{\frac{1}{2}}^m)$, and an arithmetic unit characterised by the parameters $(r_\infty^a, n_{\frac{1}{2}}^a)$. If the computational intensity is f then we transfer a vector of length n from memory and perform f, vector operations upon it. If the combined operation of the two pipelines is expressed by the parameters $(r_\infty, n_{\frac{1}{2}})$ and we assume that there is *no overlap* between the memory access and the arithmetic, then we can add the time for data transfer and the time for arithmetic to obtain (see Hockney and Jeshope [54], page 106-107)

$$r_\infty = \frac{\hat{r}_\infty}{(1 + f_{\frac{1}{2}}/f)} = \hat{r}_\infty \, \text{pipe}(f/f_{\frac{1}{2}}) \tag{3.32}$$

$$n_{\frac{1}{2}} = \frac{n_{\frac{1}{2}}^a + n_{\frac{1}{2}}^m f_{\frac{1}{2}}/f}{(1 + f_{\frac{1}{2}}/f)} = n_{\frac{1}{2}}^a \, \text{pipe}(f/f_{\frac{1}{2}}) + n_{\frac{1}{2}}^m \, \text{pipe}(f_{\frac{1}{2}}/f) \tag{3.33}$$

where the peak performance is symbolised by $\hat{r}_\infty = r_\infty^a$, and $f_{\frac{1}{2}} = (r_\infty^a / r_\infty^m)$ is called the half-performance intensity in analogy with the definition of $n_{\frac{1}{2}}$. In the above notation, the subscript infinity denotes infinite vector length, and the hat denotes infinite computational intensity. We note the occurrence of the pipeline function again.

Equation (3.32) shows that the asymptotic performance, r_∞, increases towards the peak, $\hat{r}_\infty$, as the computational intensity increases, because as this becomes larger, the effects of memory access or communication delays become negligible compared to the time spent on arithmetic. In analogy with $n_{\frac{1}{2}}$, the *half-performance intensity*, $f_{\frac{1}{2}}$ is the computational intensity required to achieve half this peak. The parameter $f_{\frac{1}{2}}$, like $n_{\frac{1}{2}}$, measures an unwanted overhead and should be as small as possible. For a full discussion of this model, including the effect of overlapping communication and arithmetic, see Hockney and Curington [53].

We see from Eqn.(3.32) that the required computer parameter to quantify memory (or communication) bottleneck effects is the half-performance intensity or $f_{\frac{1}{2}}$. If memory access and arithmetic are not overlapped as assumed above, then $f_{\frac{1}{2}}$ is the ratio of arithmetic speed (in Mflop/s) to memory access speed (in Mw/s) [54]. That is to say, it is how much faster the arithmetic is than the memory. This is the computer parameter with which the program parameter f must be compared, in order to assess the extent of performance degradation that is due to inadequate memory bandwidth. The peak arithmetic performance is a measured quantity which should be close to the theoretical peak performance r^* of Eqn.(3.1). If it is not, it suggests that there is something very wrong with the computer architecture or compiler optimisation.

The realised (or sustained) performance on long vectors is r_∞, and Eqn.(3.32) shows how this can be calculated for any value of computational intensity. If the performance on a finite length vector is required, then Eqn.(3.5), which gives the degradation in performance due to inadequate vector length, must also be used.

The variation of computer performance with f, the ratio of arithmetic to memory references, suggests that problems might usefully be classified according to their value of computational intensity. The following cases can immediately be identified:

(1) Vector Problems: $f = O(1)$			
	Dyadic Operation	$a = b \times c$	$f = \frac{1}{3}$
	All Vector Triad	$a = b + c \times d$	$f = \frac{1}{2}$
	DAXPY or level-1 BLAS	$a = \alpha b + c$	$f = \frac{2}{3}$
	Tridiagonal Solve	$[a, b, c]x = b$	
		$flop = 5n;\ \ mref = 5n$	$f = 1$
	Level-2 BLAS	$a = a + \alpha A b$	$f = 2$
(2) Logarithmic Problems: $f = O(\log n)$			
	FFT	$flop = 2\frac{1}{2} n \log_2 n$	
		$mref = 2n$	$f = 1\frac{1}{4} \log_2 n$
(3) Matrix Problems: $f = O(n)$			
	Full Matrix Solve	Ax=b	
		$flop = \frac{1}{3} n^3;\ \ mref = 2n^2$	$f = \frac{1}{3} n$
	Matrix Multiply	$A = B \times C$	
		$flop = 2n^3;\ \ mref = 3n^2$	$f = \frac{2}{3} n$
	Level-3 BLAS		$f = \frac{1}{2} n$

Where capital letters are matrices, small letters are vectors and Greek letters are scalars.

3.6.1 The POLY Benchmarks for ($\hat{r}_\infty$,$f_{\frac{1}{2}}$)

The POLY benchmarks are designed to measure the above memory-bottleneck and communication-bottleneck parameters ($\hat{r}_\infty$,$f_{\frac{1}{2}}$) for three cases of data in-cache (POLY1), data out-of-cache (POLY2), and data on another processor (POLY3). In order to reach as close to the theoretical peak performance as possible we choose the vector evaluation of a polynomial by Horner's rule as the computational kernel within the skeleton code of Fig. 3.3. This allows the parallel use of the floating-point multiplier and adder in computers that allow this. For example, if we take a third-degree polynomial, then the kernel is

```
      DO 10 I=1,N
10    Y(I)=S0+X(I)*(S1+X(I)*(S2+X(I)*S3))
```

In this case the computational intensity can be seen to be three, and is in general the order of the polynomial. Note that we do not count the scalars $S0, S1$ etc. as requiring memory references because they are assumed to have been prefetched and therefore available in registers. To measure $f_{\frac{1}{2}}$, an outer loop is added to the kernel loop to increase the order of the polynomial from one to ten, and the measured performance for long vectors, r_∞, is recorded for each value of f and fitted to Eqn.(3.32). The values of $f_{\frac{1}{2}}$ and $\hat{r}_\infty$ are obtained by considering (f/r_∞) as a function of f, and fitting the best straight line by least-squares. From Eqn.(3.32) we have

$$\frac{f}{r_\infty} = \hat{r}_\infty^{-1}(f + f_{\frac{1}{2}}) \tag{3.34}$$

Then by analogy with Eqn.(3.3), we see that $\hat{r}_\infty$ is the inverse slope of the straight line and $f_{\frac{1}{2}}$ is the negative intercept of the line with the f-axis.

The POLY1 benchmark repeats the polynomial evaluation for each order typically 1000 times for vector lengths up to 10 000, which would normally fit into the cache of a cache-based processor. Except for the first evaluation the data will therefore be found in the cache. POLY1 is therefore an *in-cache* test of the memory bottleneck between the arithmetic registers of the processor and its cache.

POLY2, on the other hand, flushes the cache prior to each different order and then performs only one polynomial evaluation, for vector lengths from 10 000 up to 100 000, which would normally exceed the cache size. Data will have to be brought from off-chip memory, and POLY2 is an *out-of-cache* test of the memory bottleneck between off-chip memory and the arithmetic registers.

3.6.2 Running POLY1 and POLY2

To compile and link the benchmark type: '`make`'. POLY1 and POLY2 are a single processor benchmarks, and on some systems it may be necessary to allocate a processor before running the benchmarks, eg. on the iPSC/860 to reserve a single processor,

type: `getcube -t1`.

To run the benchmark type: `poly1` or `poly2`

In POLY1, if the timing results are too inaccurate the parameter `NITER` in file '`poly1.inc`' may be increased. This is the number of repetitions of the kernel loop used to extend the length of time measured. `NITER=1000` is a sensible starting value. `NITER=10` may be used for testing execution but is probably too small for accurate timing.

The order of execution of the nested `DO` loop should be as specified in the Fortran code (in `SUBROUTINE DOALL`). Nonsense results (e.g. negative $f_{\frac{1}{2}}$) may be produced if the compiler tampers with the loop ordering or does software pipelining. The polynomial must be completely evaluated for one value of the loop index-I (e.g. verb+DO 310+ loop) before the next value of `I` is taken.

Output from the benchmark is written to the file '`poly1.res`' or '`poly2.res`', which should be saved before it is overwritten.

POLY2 uses vector lengths that are intended to exceed the cache size, and a single executing of the kernel is made. $f_{\frac{1}{2}}$ is then a measure of the ratio arithmetic performance (Mflop/s) to out-of-cache (i.e. main) memory access rate (Mw/s). For this reason the parameter `NITER` is not operative and is fixed at 1. In all other respects POLY2 operates like POLY1.

3.6.3 Communication Bottleneck: POLY3

POLY3 assesses the severity of the communication bottleneck. It is the same as the POLY1 benchmark except that the data for the polynomial evaluation is stored on a neighbouring processor. The value of $f_{\frac{1}{2}}$ obtained therefore measures the ratio of arithmetic to communication performance. Equation (3.32) shows that the computational intensity of the calculation must be significantly greater than $f_{\frac{1}{2}}$ (say 4 times greater) if communication is not to be a bottleneck. In this case the computational intensity is the ratio of arithmetic performed on a processor to words transferred to/from it over communication links.

In the solution method called *domain decomposition*, a problem is divided up into regions which communicate data across their common surfaces, but do calculations throughout the volume of their interiors. In this case when the amount of arithmetic is proportional to the volume of a region and the data communicated is proportional to its surface, the computational intensity is increased as the size of the region (or granularity of the decomposition) is increased. For example, a cubical region of side a has a volume to surface ratio of $\frac{1}{6}a$, so that f will be proportional to the length of the side of the cube. Then the $f_{\frac{1}{2}}$ obtained from this benchmark is directly related to the granularity that is required to make communication time unimportant. One might say, for the example above, that the side of the cube should be at least as big as to make $f > 4f_{\frac{1}{2}}$.

3.6.4 Running POLY3

POLY3 is controlled by the data file '`poly3.dat`' which contains two lines. As with COMMS1, the first specifies **`NNODE`**, the number of nodes assigned to the test, and the second **`NSLAVE`** gives the slave processor number in the range `1` to **`NNODE-1`** with which the master node (numbered 0) is communicating.

To compile and link the benchmark type: '`make`' . On some systems it may be necessary to allocate the appropriate resources before running the benchmark. For example on the iPSC/860, if **`NNODE=32`** then one must reserve 32 processors,

by typing: `getcube -t32`.

To run the benchmark type: `poly3`

Output from the benchmark is written to the file '`poly3.res`'.

`NITER` in the file '`poly3.inc`' can be varied to alter the number of repeats made, and increase the accuracy of the time measurement. Values of 100 or 1000 would be usual when taking measurements. Values of 1 or 10 might be used for short runs to test execution, but are probably too small for satisfactory timing.

As with the other POLY benchmarks, the order of executing of the kernel loop should be as specified in the Fortran code.

3.6.5 Example Results for POLY benchmarks

Figure 3.12 shows the first page of output for the POLY1 benchmark run on scalar nodes of the Meiko CS-2. This contains the introductory specification of the computer system and the software used, followed by the detailed results for the measurement of r_∞ for a computational intensity, f, of unity. As with the previous output results, the time is given for a series of vector (or kernel-loop lengths) together with the (r_∞,$n_{\frac{1}{2}}$) values of the straight-line fit and the error of the fit. In addition a final column gives the average performance for each loop length (i.e. the flop computed for this particular loop-length divided by the time). We would expect the average performance values to approach the value of r_∞ as the loop length increases, and indeed this is the case.

In the full output, several pages follow which contain similar data for values of f from 2 to 9, and finally in Fig. 3.13 we show the final page with the detailed output for $f = 10$ followed by the calculation of $f_{\frac{1}{2}}$. The latter records the last values of r_∞ obtained for each value of f, the values of ($\hat{r}_\infty$,$f_{\frac{1}{2}}$) obtained by fitting Eqn.(3.32) to this data, and the error of the fit. The second column shows clearly the increase of r_∞ as f increases, reaching half its final value of $r_\infty = 8.53$ for f between 1 and 2. This is confirmed by the final value obtained of $f_{\frac{1}{2}} = 1.74$.

```
=================================================
===  GENESIS / PARKBENCH Parallel Benchmarks  ===
===                                           ===
===                  POLY1                    ===
===                                           ===
===   Program:  In-cache (r-hat f-half)       ===
===   Version:  Standard Fortran 77           ===
===   Author:   Roger Hockney                 ===
===   Update:   November 1993; Release: 1.0   ===
=================================================
Run on Meiko CS-2 at the University of Southampton
Compiler version: SC2.0.1 20 Apr 1993 Sun FORTRAN
2.0.1 patch 100963-03, b/end SC2.0.1 03 Sep 1992
Operating system: SunOS 5.1 MEIKO FCS

COMPUTATIONAL INTENSITY = flop per mem ref =  1
----------------------------------------
Floating operations per iteration =   2
Memory references   per iteration =   2
```

LOOP LENGTH	LOOP SECS	RINF MFLOP/S	N1/2 VEC.LEN	ERROR %	R-AVERAGE MFLOP/S
1	8.709E-07	0.000000	0.000	0.00	2.296531
5	3.065E-06	3.646708	0.588	0.00	3.263028
10	5.802E-06	3.650091	0.591	0.16	3.446845
20	1.127E-05	3.654260	0.596	0.21	3.548986
50	2.773E-05	3.648195	0.583	0.24	3.605597
100	5.510E-05	3.650612	0.592	0.16	3.629580
200	1.099E-04	3.651974	0.601	0.10	3.641324
400	2.193E-04	3.631120	0.522	3.96	3.647213
600	3.286E-04	3.647438	0.887	2.84	3.651851
800	4.388E-04	3.649024	0.935	2.03	3.645952
1000	5.487E-04	3.648682	0.922	1.56	3.644995

Figure 3.12: Detailed output for the POLY1 in-cache memory-bottleneck benchmark for the Meiko CS-2

The straight line fit, and therefore Eqn.(3.32), is obviously almost exact in this case, because of the small percentages recorded in the error column and the almost constant values for $\hat{r}_\infty$ and $f_{\frac{1}{2}}$ for all values of f.

Figure 3.14 shows just the 'Calculation of $f_{\frac{1}{2}}$' data for the POLY2 'out-of-cache' benchmark on the Meiko CS-2. The fit to the computational model of Eqn.(3.32) is equally good in this case, and leads to a value of $\hat{r}_\infty$ the same within one percent as was obtained for the POLY1 'in-cache' measurement, which is to be expected. However the value of $f_{\frac{1}{2}}$ has increased from 1.7 flop/mref for POLY1 to 2.4 flop/mref in POLY2, reflecting the additional delay associated with getting data to and from the off-chip memory.

Finally the results for POLY3 'inter-processor' communication-bottleneck test are given in Fig. 3.15. For this measurement, the range of computational intensity is increased to values up to 1000 flop/mref, because of the much larger communication delays compared to memory access delays in the same processor (whether to cache or to off-chip memory). The linear fit is not quite as clean as for the same-processor tests, but the values of $\hat{r}_\infty = 7.5$Mflop/s and $f_{\frac{1}{2}} = 31$ flop/mref have stabilised well over the last five values of f. In this case it is more easy

```
COMPUTATIONAL INTENSITY = flop per mem ref =  10
-----------------------------------------
Floating operations per iteration =  20
Memory references   per iteration =   2

 LOOP      LOOP        RINF        N1/2    ERROR  R-AVERAGE
LENGTH     SECS       MFLOP/S    VEC.LEN     %     MFLOP/S
   1    2.659E-06   0.000000    0.000     0.00    7.520698
   5    1.204E-05   8.526792    0.134     0.00    8.304600
  10    2.369E-05   8.560406    0.144     0.71    8.442737
  50    1.174E-04   8.536902    0.134     0.26    8.514711
 100    2.363E-04   8.477657    0.039     1.26    8.463358
 200    4.703E-04   8.504605    0.115     0.83    8.505513
 400    9.367E-04   8.541727    0.289     0.81    8.540497
 600    1.405E-03   8.544082    0.311     0.52    8.541057
 800    1.873E-03   8.544590    0.318     0.37    8.541577
1000    2.347E-03   8.534130    0.147     0.55    8.522882

              CALCULATION OF FHALF
              --------------------

  F        RINF          RHAT        FHALF     ERROR
 f/m      Mflop/s       Mflop/s    flop/mref     %
 1.0     3.648682      0.000000     0.0000     0.00
 2.0     5.349724     10.022106     1.7468     0.00
 3.0     6.335930     10.029162     1.7489     0.01
 4.0     6.974400     10.020017     1.7458     0.02
 5.0     7.420364     10.008261     1.7414     0.03
 6.0     7.768743     10.025156     1.7483     0.05
 7.0     8.026592     10.030919     1.7509     0.04
 8.0     8.235885     10.036965     1.7537     0.04
 9.0     8.384456     10.023524     1.7469     0.06
10.0     8.534130     10.024444     1.7474     0.05

         LAST VALUES : RHAT =  10.024444 Mflop/s
                      FHALF =   1.7474 flop/mref
```

Figure 3.13: Summary output for the POLY1 in-cache memory-bottleneck benchmark for the Meiko CS-2, showing the calculation of $(\hat{r}_\infty, f_{\frac{1}{2}})$

to see that r_∞ reaches half of its peak value of $\hat{r}_\infty$ for f lying between 20 and 40 flop/mref, corresponding to the value of 31 flop/mref obtained for $f_{\frac{1}{2}}$. The value of $f_{\frac{1}{2}}$ means that one must perform about 120 $(4 \times f_{\frac{1}{2}})$ flop per data access to another processor to achieve 80% of $\hat{r}_\infty$ or 6 Mflop/s. This is a high requirement for the computational intensity of an algorithm.

3.7 SYNCH1 Synchronisation Benchmark

SYNCH1 measures the time to execute a barrier synchronisation statement as a function of the number of processors taking part in the barrier. A barrier statement operates as follows: all processors call the barrier statement in their programs, and no processor can proceed past the barrier in its program until *all* processors have reached the barrier. This is the most elementary form of global synchronisation in both shared-memory and distribute-memory

```
=================================================
=== GENESIS / PARKBENCH Parallel Benchmarks  ===
===                                          ===
===                 POLY2                    ===
===                                          ===
===  Program:  Out-cache (r-hat f-half)      ===
===  Version:  Standard Fortran 77           ===
===  Author:   Roger Hockney                 ===
===  Update:   November 1993; Release: 1.0   ===
=================================================
Run on Meiko CS-2 at the University of Southampton
Compiler version: SC2.0.1 20 Apr 1993 Sun FORTRAN
2.0.1 patch 100963-03, b/end SC2.0.1 03 Sep 1992
Operating system: SunOS 5.1 MEIKO FCS

            CALCULATION OF FHALF
            --------------------

  F         RINF         RHAT        FHALF     ERROR
 f/m      Mflop/s      Mflop/s    flop/mref      %
 1.0     2.897384     0.000000     0.0000     0.00
 2.0     4.675640    12.105070     3.1779     0.00
 3.0     5.740047    11.267291     2.8657     0.55
 4.0     6.317115    10.426665     2.5276     1.10
 5.0     6.752244    10.038520     2.3590     1.10
 6.0     7.330663    10.247407     2.4567     1.03
 7.0     7.468581    10.087637     2.3768     0.94
 8.0     7.831588    10.148712     2.4093     0.83
 9.0     8.002175    10.145421     2.4075     0.71
10.0     8.132376    10.107997     2.3851     0.63

    LAST VALUES :  RHAT =  10.107997 Mflop/s
                  FHALF =     2.3851 flop/mref
```

Figure 3.14: Summary output for the POLY2 out-of-cache memory-bottleneck benchmark for the Meiko CS-2, showing the calculation of $(\hat{r}_\infty, f_{\frac{1}{2}})$

parallel computers. It ensures that all work that comes before the barrier—in all the programs and on all the processors—is completed before any work is done in program statements that appear after the barrier statement in any processor.

In shared-memory programs, barriers are used to ensure that all the data computation in one stage of a parallel program is completed *before* it is used in a later stage. In the absence of the barrier it is possible, even likely, that some processors will rush ahead with the second stage before their input data has been completely computed by other processors still working on the first stage, giving rise to incorrect numbers. Futhermore, the final results will be variable from run to run because they will depend on the relative speed with which the different processors compute. In distributed-memory computers, barriers are less necessary if blocking send and receives are used because they themselves prevent data being used before it is ready. However barrier statements are provided in most distributed-memory programming systems and are used for overall synchronisation.

A barrier statement is a global operation which requires information to be sent to and from all processors taking part in the barrier. The practicability of massively parallel computation with thousands or tens of thousands of processors therefore depends on the time for a barrier

```
=====================================================
===   GENESIS/PARKBENCH Distributed Memory        ===
===              Benchmarks                       ===
===                                               ===
===                POLY3                          ===
===                                               ===
===     Program:   Communication Bottleneck       ===
===     Version:   PVM +  Fortran 77              ===
===     Author:    Roger W. Hockney               ===
===     Update:    October 1993;                  ===
=====================================================
Run on Meiko CS-2 at the University of Southampton
Compiler version: SC2.0.1 20 Apr 1993 Sun FORTRAN
2.0.1 patch 100963-03, b/end SC2.0.1 03 Sep 1992
Operating system: SunOS 5.1 MEIKO FCS

                 CALCULATION OF FHALF
                 --------------------
A straight line is being fitted to y as function of x.
Inverse slope is RHAT and negative intercept is FHALF.

  F=x       y       RINF         RHAT       FHALF    ERROR
  f/m    us/mref   Mflop/s      Mflop/s   flop/mref    %
   1.0     9.97   0.100273     0.000000     0.0000    0.00
   2.0     4.23   0.472599    -0.174191    -2.7372    0.00
   4.0     4.33   0.922437    -0.601303    -5.9652   36.17
   6.0     4.55   1.318279    -1.313383   -10.5147   35.50
   8.0     4.76   1.677247    -2.467493   -17.2931   32.74
  10.0     4.95   2.017992    -4.319069   -27.6644   30.00
  20.0     6.36   3.142246   -96.126144  -510.8903   25.01
  40.0     9.02   4.431445    11.433621    53.9770   17.86
  60.0    11.70   5.126160     9.113810    41.5112   13.44
  80.0    14.40   5.553063     8.453437    37.8320   10.61
 100.0    17.04   5.865969     8.173338    36.1989    8.70
 200.0    30.52   6.551394     7.776482    33.5489    4.76
 400.0    57.66   6.936430     7.547428    31.7563    2.48
 600.0    84.34   7.113939     7.506086    31.3746    1.65
 800.0   110.86   7.215664     7.501593    31.3250    1.22
1000.0   137.92   7.250525     7.489190    31.1607    0.96

      LAST VALUES :     RHAT =  7.489190 Mflop/s
                       FHALF = 31.1607 flop/mref
```

Figure 3.15: Summary output for the POLY3 inter-processor benchmark for PVM on the Meiko CS-2, showing the calculation of $(\hat{r}_\infty, f_{\frac{1}{2}})$

```
==================================================
===   GENESIS Distributed Memory Benchmarks      ===
===                                              ===
===                 SYNCH1                       ===
===                                              ===
===     Program:  Barrier Syncronisation Rate    ===
===     Version:  PARMACS Fortran 77             ===
===     Author:   Roger W. Hockney               ===
===     Update:   May 1993; Release: 2.2         ===
==================================================
Run on Meiko CS-2 at the University of Southampton
Compiler version: SC2.0.1 20 Apr 1993 Sun FORTRAN
2.0.1 patch 100963-03, b/end SC2.0.1 03 Sep 1992
Operating system: SunOS 5.1 MEIKO FCS
          -----------------------------------
          SYNCH1: Global Synch - Barrier Rate
          -----------------------------------
     Number of processors (nodes) =      4
     Time per barrier =     5.341E+01 us
     Barrier Rate     =     1.872E-02 Mbarr/s
```

Figure 3.16: Output for the SYNCH1 benchmark for PVM on the Meiko CS-2

not increasing too fast with the number of processors. It is intended that the benchmark be run for a sequence of different numbers of processors in order to determine this variation. The results are quoted both as a barrier time, and as the number of barrier statements executed per second (barr/s).

The SYNCH1 benchmark measures the overhead for global synchronisation by measuring the rate at which the PVM barrier statement (pvmfbarrier) can be executed, as a function of the number of processes (nodes) taking part in the global barrier synchronisation. The SYNCH1 benchmark repeats a sequence of 10 barrier statements 1000 times. Although the first release of Parkbench benchmarks measures the performance of the PVM implementation of a barrier, it is simple matter to replace this macro and measure any other implementation of a barrier. The results can then be used to compare the effectiveness of the different software systems at global synchronisation (for example, comparing PVM both general-release and native implementations with PARMACS and MPI).

3.7.1 Running SYNCH1

SYNCH1 is controlled by the data file '`synch1.dat`' which contains the single line specifying `NNODE`, the number of nodes taking part in the test.

To compile and link the benchmark type: '`make`'. On some systems it may be necessary to allocate the appropriate resources before running the benchmark, eg. on the iPSC/860 to reserve a cube of 8 processors, type: `getcube -t8`

To run the benchmark executable, type: `synch1`

This will automatically load both host and node programs. The progress of the benchmark execution can be monitored via the standard output, whilst a permanent copy of the benchmark is written to a file called '`synch1.res`'. If the run is successful and a permanent record is required, the file '`synch1.res`' should be copied to another file before the next run overwrites it. As an example of this output, Fig. 3.16 shows the results of running SYNCH1 under PARMACS on the Meiko CS-2 with four processors.

3.8 Summary of Benchmarks

Table-3.1 summarises the current low-level benchmarks, and the architectural properties and parameters that they measure.

Table 3.1: Current Low-Level benchmarks and the Parameters they measure. Note we abbreviate performance (perf.), arithmetic (arith.), communication (comms.), operations (ops.).

Benchmark	Measures	Parameters
SINGLE-PROCESSOR		
TICK1	Timer resolution	tick interval
TICK2	Timer value	wall-clock check
RINF1	Basic Arith. ops.	$(r_\infty, n_{\frac{1}{2}})$
POLY1	Cache-bottleneck	$(\hat{r}_\infty, f_{\frac{1}{2}})$
POLY2	Memory-bottleneck	$(\hat{r}_\infty, f_{\frac{1}{2}})$
MULTI-PROCESSOR		
COMMS1	Basic Message perf.	$(r_\infty, n_{\frac{1}{2}})$
COMMS2	Message exch. perf.	$(r_\infty, n_{\frac{1}{2}})$
COMMS3	Saturation Bandwidth	$(r_\infty, n_{\frac{1}{2}})$
POLY3	Comms. Bottleneck	$(\hat{r}_\infty, f_{\frac{1}{2}})$
SYNCH1	Barrier time and rate	barr/s

Chapter 4

Computational Similarity and Scaling

This chapter is concerned with understanding the scaling of parallel computer performance as a function of both problem size and the number of processors used. The presentation of the first section-4.1 was inspired by a seminar given by Horace Flatt in 1987 and developed in his paper with Ken Kennedy [28]. The rest of the chapter on computational similarity follows Hockney [51].

4.1 Basic Facts of Parallel Life

It is helpful, first, to consider the parallelisation of a constant-sized problem on a p-processor parallel computer. Suppose the execution time, T_1, of the original unparallelised code can be divided into a time, T_s, for computational and organisational work that must be executed on a single processor (called the *serial* component), and a part, T_{par}, for work that can be distributed equally amongst the p processors (we assume perfect load balancing), then

$$T_1 = T_s + T_{par} \tag{4.1}$$

It is important to realise that the component T_s includes the time for any code that is repeated on all processors for convenience, in order to avoid performing the calculations on one processor and subsequently broadcasting the results to all the others. In other words, no real parallelisation takes place if one unnecessarily replicates the work to be done by a factor p, and then subsequently divides it by p when distributing the replicated work across the processors.

After parallelisation the component T_s remains unchanged, but the time for the component T_{par} is divided by p because each processor performs $1/p$th of the work in parallel (i.e. simultaneously) with the others, in $1/p$ of the original time. Then the time for the parallelised version of the code is

$$T_p = T_s + \frac{T_{par}}{p} + T_{sc}(p) \tag{4.2}$$

where $T_{sc}(p)$ is the extra time introduced into the parallelised code for synchronising the processors. In distributed-memory systems it also includes the communication time required to move data into the correct processor before computation. These are both activities that are not necessary if the code is run on a single processor, and therefore $T_{sc}(p)$ constitutes the overhead of going parallel. Furthermore $T_{sc}(p)$ will be a monotonically increasing function of the number of processors for any problems that require global synchronisation (e.g. tests for convergence) or global communication (e.g. as in the FFT).

The appropriate dimensionless performance metric to use in the study of scaling turns out to be the conventional Speedup, or ratio of one-processor time to p-processor time (see section-2.5 and 4.2.2). Both Snelling and the author [42, 43, 46, 49] have discussed extensively the *invalidity* of using Speedup as though it were an absolute measure of parallel performance suitable for comparing the performance of different computers (see section-2.6). However, its use in this context as a relative measure, and a natural definition of dimensionless performance, is perfectly valid.

The ideal situation for parallelisation is when $T_s = T_{sc} = 0$ or are negligible compared to T_{par}/p. In this case the Speedup is

$$S_p = \frac{T_1}{T_p} = p \tag{4.3}$$

and the Speedup increases linearly with p. This is called the ideal linear Speedup relationship, and gives the promise of unlimited increase in computer performance as more processors are added to the system.

If, however, there is a serial component ($T_s \neq 0$), but still synchronisation and communication time can be ignored ($T_{sc} = 0$), then

$$S_p = \frac{T_1}{(T_s + T_{par}/p)} = \frac{S_\infty}{(1 + p_{\frac{1}{2}}/p)} = S_\infty \mathrm{pipe}(p/p_{\frac{1}{2}}) \tag{4.4}$$

where $S_\infty = T_1/T_s$ and $p_{\frac{1}{2}} = T_{par}/T_s$, from which we obtain the identity $S_\infty = 1 + p_{\frac{1}{2}}$. This identity is a consequence of the requirement that the Speedup for one processor is unity.

Thus we find that the Speedup saturates at a maximum S_∞ as the number of processors increases, and cannot exceed a value equal to the inverse of the fraction of the original code time that cannot be parallelised. This saturation in Speedup due to an inherently serial or sequential part that must be executed on a single processor, is called the Amdahl saturation effect, and can be characterised by the parameter pair (S_∞,$p_{\frac{1}{2}}$) and the familiar pipeline function. The rapidity with which the asymptotic value is reached is determined by $p_{\frac{1}{2}}$ which is the ratio of the parallelisable time to the non-parallelisable time in the original code. The parameters are properties of the algorithm being parallelised, and clearly there is little point in using more than, say, $4p_{\frac{1}{2}}$ processors to execute such an algorithm, because 80% of the maximum possible Speedup will have been gained. The addition of more processors would only make a marginal further improvement in performance, that is unlikely to be cost effective.

Any algorithm of fixed size with a non-zero serial component, has a finite amount of usable parallelism within it, and can only make effective use of a finite number of processors on a multiprocessor system. We regard the algorithmic parameter $p_{\frac{1}{2}}$ as measuring this algorithmic parallelism. Whilst it is true that the algorithmic parallelism, can be increased by increasing the problem size, this is not a useful option if the desire is to use more processors to decrease the execution time of a problem that is already deemed to be large enough to perform the task in hand with sufficient accuracy.

If, next, we consider that the act of going parallel introduces additional overheads $T_{sc}(p)$ into the original program, and that these overheads are likely to increase with p, then the Speedup will rise to a maximum, $\tilde{S}_p$, at say $\tilde{p}$ processors, and for $p > \tilde{p}$ processors, the performance will decrease (rather than increase) as more processors are used. There are many examples of such maxima in Speedup curves. The peak therefore represents an optimum in performance and optimum number of processors, and this is used in the subsequent analysis to simplify the understanding of scaling. In fact, Flatt and Kennedy [28] show that under

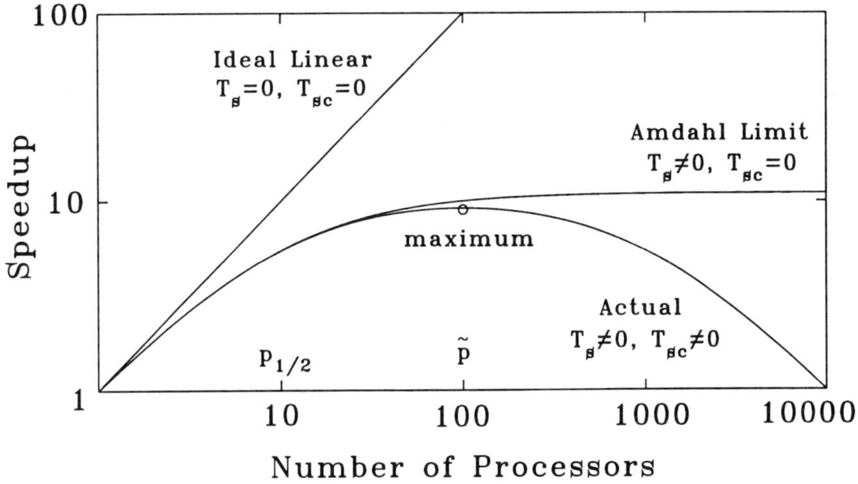

Figure 4.1: Figure showing ideal linear Speedup, Amdahl saturation and the 'Actual' Speedup with a maximum in performance for the constructed example of Eqn. (4.6)

reasonable assumptions about the cost of synchronisation, there *will exist a unique minimum* in time, and therefore a *unique maximum* in performance (their theorem 3.4).

The above three cases are illustrated in Fig. 4.1 for a constructed example in which the parallelised time is given by

$$T_p = 1 + \frac{10}{p} + 0.001p \tag{4.5}$$

in which $T_s = 1$, $T_{par} = 10$ and $T_{sc} = 0.001p$. This corresponds to a code in which 91% of the original time is successfully parallelised (leaving 9% unparallelised), and the synchronisation time (last term) is a linear function of p. Ignoring synchronisation, the first two terms give Amdahl saturation with $p_{\frac{1}{2}} = 10$ and a maximum Speedup of $S_\infty = 11$ (see the curve marked 'Amdahl Limit').

The small factor multiplying p in the synchronisation term means that this overhead is negligible (less than 10%) for less than 10 processors, but the proportionality with p ensures that eventually, with p greater than 1000, synchronisation dominates the calculation time. This causes there to be a maximum in the performance at $p = \tilde{p}$ which is shown in the curve labelled 'Actual'.

A little algebra shows that the Speedup for a linear synchronisation term ($T_{sc} = bp$) can be expressed, in general, as

$$S_p = \frac{S_\infty}{\left[1 + \frac{p_{\frac{1}{2}}}{p}\left\{1 + \left(\frac{p}{\tilde{p}}\right)^2\right\}\right]} \tag{4.6}$$

where $\tilde{p} = \sqrt{(T_{par}/b)}$. In this example $\tilde{p} = 100$, and the maximum is shown as a circle in Fig.4.1. Differentiating Eqn. (4.6) with respect to p, and setting this derivative to zero, shows that the maximum occurs when $p = \tilde{p}$. Setting $p = \tilde{p}$ in Eqn. (4.6) gives the Speedup at the maximum as

$$\tilde{S}_p = \frac{S_\infty}{\left[1 + 2\left(\frac{p_{\frac{1}{2}}}{\tilde{p}}\right)\right]} \tag{4.7}$$

Thus in this example $\tilde{S}_p = 9.17$.

A generalisation of the above synchronisation model to higher powers of p is given by the author in reference [44] and [54] page 115. If the synchronisation term is $T_{sc} \propto p^{n-1}$ where n is called the index of synchronisation, then any of the inverse-time performance metrics of Chapter-2, $R(p)$, can be expressed

$$R(p) = \frac{R_\infty}{\left[1 + \frac{p_{\frac{1}{2}}}{p}\left\{1 + \frac{1}{(n-1)}\left(\frac{p}{\tilde{p}}\right)^n\right\}\right]} \tag{4.8}$$

The maximum performance is still at $p = \tilde{p}$, but the performance at the maximum is given by

$$\tilde{R} = \frac{R_\infty}{\left[1 + \frac{n}{(n-1)}\left(\frac{p_{\frac{1}{2}}}{\tilde{p}}\right)\right]} \tag{4.9}$$

Temporal performance data for a parallelised Particle-In-Cell (PIC) code on the first IBM parallel computer (Enrico Clementi's LCAP) fitted this model very well with $R_\infty = 0.5$ tstep/s, $p_{\frac{1}{2}} = 5.5$, $\tilde{R} = 0.25$ tstep/s, $\tilde{p} = 7.1$ and $n = 5$.

4.2 Introducing the DUSD Method

One of the principal problems of parallel computation is understanding how the performance of a program or benchmark varies with the number of processors and with the problem size. We are also keenly interested to know how to extrapolate results from one computer to another computer with quite different hardware characteristics. Perhaps this is impossible, but it might be that there are some dimensionless ratios of hardware and program parameters that determine whether the computing conditions are similar in the two cases, and that, if so, a similar performance might be expected.

In 1883 Osborne Reynolds [63] showed how plotting experimental results in terms of dimensionless quantities greatly simplified their interpretation, to the extent that previously disparate experimental curves were shown to follow approximately the same line when expressed as appropriate dimensionless ratios. In the study of fluid flow at different velocities through pipes of different radii, he discovered that a certain dimensionless ratio, now called the Reynolds number, determines the character of the flow (whether it is laminar or turbulent). This pure number is a combination of a parameter describing the pipe (its radius), and parameters describing the fluid (its density, speed and viscosity). Turbulence appears at a certain critical value of Reynolds number, and two flows are dynamically similar if they have the same Reynolds number, even though the values of the individual parameters entering into the Reynolds number may be quite different. This similarity is widely used in scaling experimental results obtained by ship and aircraft models to full-size vehicles.

Curiously, such dimensional analysis, although widely used successfully in other branches of science and engineering, has scarcely been used in computer science. Notable exceptions are the use of dimensionless hardware ratios by Gropp and Keyes in the analysis of the scaling of domain decomposition methods [34]; and the work of Numrich on memory contention in shared-memory multiprocessors [60, 61] and the scaling of communication rates in distributed-memory computers [62]. Our work extends the above by introducing the effect of problem size into the dimensionless parameters, and eliminating the number of processors as an independent variable.

In this chapter we try to identify dimensionless quantities that play a similar role in computer performance analysis as do Reynolds and other dimensionless numbers in fluid dynamics.

We start with a three-parameter description of the computer hardware, and a timing rel tion using these parameters which approximately describes the performance of a class of computer programs, including the Genesis FFT1 benchmark [36, 35]. The timing relation thereby defines the class of programs that are being considered. If we express the absolute performance as the product of the single-processor performance times the traditional speedup, here called the self-speedup (see equation 4.31), we find that the optimum self-speedup is a function of only two dimensionless quantities. We can therefore plot a Dimensionless Universal Scaling Diagram (DUSD, pronounced 'dusdee') which gives contours of constant value of optimum self-speedup for any value of the two dimensionless ratios. This diagram is universal in the sense that it describes the scaling of all programs or benchmarks within the defined class, and to all computers that can be described adequately by the three hardware parameters. The diagram also gives the optimum number of processors to be used to obtain the optimum self-speedup or best performance.

In analogy with fluid dynamics, two computer calculations which have the same values for the dimensionless ratios are said to be 'computationally similar', and will have consequently the same optimum number of processors and the same optimum self-speedup, even though their individual hardware and software parameters are widely different. Within the limitation of the three-parameter hardware model, this DUSD completely describes the scaling properties of the above class of programs, for all problem sizes and for all computers. There is nothing more that can, or need, be said about the matter of scaling.

Other classes of programs and benchmarks which are defined by different functional forms for the dependence of their timing relation on the number of processors, will have different DUSDs, but it is possible that a library of ten or less DUSDs might cover most of the commonly met functional forms for the timing relation. The identification of these functional forms, and the publishing of their corresponding DUSDs will go a long way to providing an understanding of the scaling of parallel programs.

The DUSD can also be used to see the effect on performance of making changes to the hardware parameters. For example, in the DUSD diagram given here, a reduction in the message latency means a vertical upward movement, whereas an increase in asymptotic bandwidth is a movement to the right. The effect of an increase in problem size can also be seen, because this means a movement upward and slightly to the right. Starting from the point representing the current hardware, this feature of the diagram can be used to identify which hardware parameter it would be most beneficial to improve.

4.2.1 The DUSD Method in General

In its simplest form the DUSD method is based on a 3-parameter timing model. Such a model contains three hardware parameters describing the computer $(r_\infty^s, r_\infty^c, t_0^c)$, and a corresponding set of three program parameters (operation or instruction counts) describing the application or benchmark program being timed $(s^s(N;p), s^c(N;p), q^c(N;p))$. If the problem size is denoted by N (possibly a collection of variables) and the number of processors by p, the total wall-clock time to completion of the program can be expressed as follows, where the three terms on the right-hand side arise respectively from the time to perform the s^s floating-point operations at the constant scalar rate of r_∞^s, the stream-time which is the time to transfer all necessary words, s^c, at the asymptotic communication bandwidth of r_∞^c, and the last term that is the total time spent in message startups, that is to say the total number of messages sent, q^c, multiplied by the message startup time, t_0^c.

$$T(N;p) = \frac{s^s(N;p)}{r_\infty^s} + \frac{s^c(N;p)}{r_\infty^c} + t_0^c q^c(N;p) \qquad (4.10)$$

Alternatively, the startup term can be expressed in terms of the message length for half performance, $n^c_{\frac{1}{2}}$, and the average message length $n^c = s^c(N;p)/q^c(N;p)$:

$$T(N;p) = \frac{s^s(N;p)}{r^s_\infty} + \frac{s^c(N;p)}{r^c_\infty}(1 + n^c_{\frac{1}{2}}/n^c) \tag{4.11}$$

where $n^c_{\frac{1}{2}} = t^c_0 r^c_\infty$.

In the above we are using the Getov/Hockney notation [30, 32, 33] in which the functions $s(N;p)$ denote a total operation count (floating-point operations, or words or bytes transferred) and $q(N;p)$ denotes the number of instructions (scalar or vector arithmetic instructions, or message sends), over which the operations are distributed. The type of instruction is specified by the superscript (s for scalar arithmetic, v for vector arithmetic, and c for communication). A semicolon is used to separate the problem size parameter N from the number of processors p, because N itself may be an abbreviation for a collection of size parameters separated by commas. Furthermore we wish to distinguish more strongly these size variables which are properties of the problem, from p which is a property of the computer and plays the main role in the analysis below.

The above timing model is also special to so-called 'uniformly distributed' applications in which the same program executes on all processors and it is assumed that the work can be uniformly distributed across the processors (i.e. the load balance is perfect) [32]. Furthermore we make the worst-case assumption that there is no overlap between arithmetic and communication, or between communication latency and stream time. In this case the arithmetic time, the message stream time and the message latency can be added, as is done in equation (4.10). If there is overlap between these times, we can still use the non-overlapping model, but will find that the measured performance will correspond to improved values for the three hardware parameters (higher arithmetic rate and communication bandwidth, and less message latency). Although the above assumptions may seem severe, they do apply to many important applications, particularly simple kernels like the FFT. It is important to realise, however, that the DUSD analysis below does not depend on these assumptions, and may be applied to more general timing relations.

Wolton (private communication) has also correctly pointed out that the values that have to be used for the three 'so-called' hardware parameters above, $(r^s_\infty, r^c_\infty, t^c_0)$, depend also on the software being used (e.g. the quality of the Fortran compiler in producing efficient object code) and on the efficiency of the particular implementation of an algorithm (e.g. a better use of cache will raise r^s_∞, even when the amount of arithmetic performed is the same). Not withstanding this caveat, we will refer to the above three parameters simply as hardware parameters in the rest of this chapter, because they are the representation of the hardware in the computational model being used which is entirely defined by either timing equation (4.10) or equation (4.11).

The first step of the DUSD method is to eliminate the dependence on p by finding, for each problem size N, the number of processors that gives the best or optimum performance. We will then assume that the parallel processor is always used with this optimum number of processors, and see what conclusions can be drawn about performance and its scaling. This is achieved by separating out the p dependence and expressing all factors like $s(N;p)$ as a product of an N-dependent factor $s_N(N)$ and a p-dependent factor $s_p(p)$, the subscripts being necessary to distinguish the functions when numerical arguments are used. Then for all superscripts $\alpha = s, v, c$

$$\begin{aligned} s^\alpha(N;p) &= s^\alpha_N(N)s^\alpha_p(p) \\ q^\alpha(N;p) &= q^\alpha_N(N)q^\alpha_p(p) \end{aligned} \tag{4.12}$$

and

$$T(N;p) = \frac{s_N^s(N)s_p^s(p)}{r_\infty^s} + \frac{s_N^c(N)s_p^c(p)}{r_\infty^c} + t_0^c q_N^c(N) q_p^c(p) \tag{4.13}$$

Given a formula for the wall-clock time of execution, the Temporal performance can be defined (see section-2.5.1):

$$R_T(N) = 1/T(N;p) \tag{4.14}$$

This is the simplest absolute measure of performance that has the property that the maximum performance corresponds to the minimum in the time. We take it as axiomatic in this context that the purpose of parallel computing is to minimise the wall-clock time for a computation, that is to say to maximise the Temporal performance. For a constant size problem (N constant), the Temporal performance normally rises to a peak as the number of processors is increased, and then subsequently decreases (see section-4.1). The existence of this peak in performance is central to the following analysis, because we are going to limit consideration to this point. In order to emphasise this, we use the tilde accent an any variable to indicate its value at the peak. The optimum number of processors to use, $\tilde{p}$, and the optimum performance, $\tilde{R}$, occur at this peak, when the time of execution is least. At this point $\partial T(N;p)/\partial p = 0$, which condition gives the optimum equation (OE) that can be solved for $\tilde{p}$:

$$\left.\frac{\partial T(N;p)}{\partial p}\right|_{p=\tilde{p}} = \frac{s_N^s(N)}{r_\infty^s} \left.\frac{ds_p^s(p)}{dp}\right|_{p=\tilde{p}} + \frac{s_N^c(N)}{r_\infty^c} \left.\frac{ds_p^c(p)}{dp}\right|_{p=\tilde{p}} + t_0^c q_N^c(N) \left.\frac{dq_p^c(p)}{dp}\right|_{p=\tilde{p}} = 0 \tag{4.15}$$

It is always necessary to examine the solution of this equation carefully, particularly if there are multiple solutions. This is done by examining the original timing equation (4.10) carefully to ensure that one picks out the solution that corresponds to the required minimum in time, as opposed to solutions that may correspond to unwanted subsidiary local minima or maximum in time. Then, for example, the optimum Temporal performance is given by

$$\tilde{R}_T(N) = 1/\tilde{T}(N;\tilde{p}(N)) \tag{4.16}$$

Note that the optimum number of processors is a function only of the problem size, and that therefore the optimum performance $\tilde{R}_T$ is also only a function of the problem size. Thus, if we agree to study scaling at this optimum point, one variable, namely p, has been removed from the problem. This is an essential simplification of the DUSD method.

Sometimes the performance does not show a peak as p increases, but alternatively shows a monotonic climb towards a theoretical saturation value, as is typical of pure Amdahl saturation. We can say that such a situation corresponds to a peak in performance at $p = \infty$, which of course can never be reached. We can use this saturation value as the optimum performance for each value of N, on the assumption that we have enough processors to get close to the saturation value. Alternatively, and more reasonably, we can define the optimum number of processors as a function of problem size, N, by defining $\tilde{p}$ as the number of processors needed to reach a specified fraction, say 80%, of the theoretical saturation value. The most general conclusions can be drawn if we assume that there are always enough processors in the system to reach $p = \tilde{p}$. If, however, the analysis is being performed for a computer system with a fixed maximum number of processors, p_{max}, then the subsequent analysis can still be performed using p_{max} in place of $\tilde{p}$ when $\tilde{p}$ exceeds p_{max}.

In the analysis above certain mathematical assumptions have been made which should be explicitly declared:

1. The existence of the peak in performance is assumed, although ways of avoiding the problem in the case of an asymptote are given.

2. The number of processors is treated as a continuous variable, whereas in reality it is discrete. However with parallel computers now sporting a thousand processors or more, this distinction is becoming less and less important.

3. The timing relation is assumed to be differentiable. This is not an essential assumption because numerical evaluation could be used to find the peak. However it is a convenient assumption, and allows a formula to be given for the optimum equation. This assumption is realistic because many timing formulae in the literature are differentiable.

4. Each term in the timing equation is assumed to be separable (or factorisable) into the product of a function of N and a function of p. This assumption is necessary for the development of the theory and reduces the number of parameters by one. It is true for many timing relations, and others could be approximated by relations satisfying this condition. Note that we do not require the whole timing relation to be separable, but only that the timing relation be a sum of separable terms.

4.2.2 Dimensional Analysis

The first step in dimensional analysis [19, 18] is to define a consistent set of units for all variables. In our notation, all parameters named δ are dimensionless, and their numbering by a subscript is arbitrary. The number of message send instructions q_N^c, the number of processors p and the problem size N are also considered to be dimensionless, the latter two because they frequently appear within logarithms. Functions of p are also dimensionless. Using the units defined in Chapter-2, the other quantities have dimensions as shown between the square brackets [ ].

$$N, p, s_p^s(p), s_p^c(p), q^c(N;p), q_N^c(N), q_p^c(p) \qquad [1] \qquad (4.17)$$

$$r_\infty^s \qquad [\text{flop/s}] \qquad (4.18)$$

$$r_\infty^c \qquad [\text{w/s}] \qquad (4.19)$$

$$t_0^c \qquad [\text{s}] \qquad (4.20)$$

$$s^s(N;p), s_N^s(N) \qquad [\text{flop}] \qquad (4.21)$$

$$s^c(N;p), s_N^c(N) \qquad [\text{w}] \qquad (4.22)$$

There is also a notation for the combinations of computer hardware parameters that appear in the dimensionless ratios

$$n_{\frac{1}{2}}^c = t_0^c r_\infty^c \qquad [\text{w}] \qquad (4.23)$$

$$n_{\frac{1}{2}}^{cs} = t_0^c r_\infty^s \qquad [\text{flop}] \qquad (4.24)$$

The first of these, $n_{\frac{1}{2}}^c$, is the ordinary half-performance length for message sending, that is to say the message length required to achieve half of the asymptotic bandwidth r_∞^c [46, 49]. It is also equal to the number of words that could have been communicated during the startup time. The second quantity, $n_{\frac{1}{2}}^{cs}$, is a kind of cross $n_{\frac{1}{2}}$ where the cost of the communication overhead t_0^c is measured in terms of how much arithmetic, rather than how much communication, could have been done during the message startup time. This quantity has already been identified as an important combination of hardware parameters by a number of researchers: Hey and Scott have referred to 'latency in terms of computational flop', and Solchenbach has identified it as one of two important combinations of hardware parameters that determine the capabilities of a parallel computer [72], and was called by him the 'latency in flop'. One is tempted to coin the word 'flopency' for this important hardware quantity.

The next step in dimensional analysis is to make all variables (including for example 'time') dimensionless by dividing them by a convenient quantity with the same dimensions. The result is a dimensionless equation in these dimensionless variables, in which all other parameters describing the original system appear as dimensionless ratios. Generally there are fewer dimensionless ratios than there were original system parameters, and the process has significantly simplified and clarified the problem.

In our case, the general timing equation (4.13) can be expressed in dimensionless form by dividing through by $t_0^c q_N^c(N)$, when two dimensionless ratios appear, δ_3 and δ_2. These two ratios contain all the dependence on problem size and computer hardware parameters. This is a significant simplification because the original dimensional timing equation (4.13) contained three hardware parameters and three functions of problem size. Thus there is nothing strange about the concept of dimensionless time (or dimensionless anything else) which we designated with a prime to the variable name. The dimensionless timing equation then becomes

$$T'(N;p) = T(N;p)/(t_0^c q_N^c(N)) = \delta_3 s_p^s(p) + \delta_2 s_p^c(p) + q_p^c(p) \tag{4.25}$$

where

$$\delta_3(N;t_0^c,r_\infty^s) = \frac{s_N^s(N)}{q_N^c(N)t_0^c r_\infty^s} \tag{4.26}$$

$$\delta_2(N;t_0^c,r_\infty^c) = \frac{s_N^c(N)}{q_N^c(N)t_0^c r_\infty^c} = \frac{n_N^c(N)}{n_{\frac{1}{2}}^c} \tag{4.27}$$

where the N dependence of the average message length is $n_N^c = s_N^c/q_N^c$. These definitions of the dimensionless δ ratios show the most useful way of making the computational work, s^s, and the average message length, n_N^c dimensionless. Hence we will refer to δ_3 as the *dimensionless work*, and to δ_2 as the *dimensionless message length*.

The dimensionless timing relation (4.25) defines the class of timing relations and corresponding programs for which the DUSD is being drawn. Even if the functional dependence of the deltas on problem size differs widely between programs through equations (4.26) and (4.27), all programs with a dimensionless timing relation with the same functional dependence on p through equation (4.25), belong to the same class and have the same DUSD.

Using the dimensionless ratios, the optimum equation (4.15) simplifies to:

$$\delta_3 \left.\frac{ds_p^s(p)}{dp}\right|_{p=\tilde{p}} + \delta_2 \left.\frac{ds_p^c(p)}{dp}\right|_{p=\tilde{p}} + \left.\frac{dq_p^c(p)}{dp}\right|_{p=\tilde{p}} = 0 \tag{4.28}$$

Since $\tilde{p}$ is the solution of this equation which contains only $\tilde{p}$ and δ_3 and δ_2, we can conclude that the optimum number of processors, $\tilde{p}$, depends only on the values of the two dimensionless ratios δ_3 and δ_2, which are certain combinations of the original three hardware parameters and the problem size. This is true for any computer which is adequately described by the above three hardware parameters, and for any problem size. Consequently, the minimum dimensionless time $\tilde{T}'$ is, from equation (4.25), also only dependent on the two dimensionless ratios. The traditional speedup (see below) is similarly dependent only on the two dimensionless ratios. The absolute execution time in seconds can be obtained by multiplying the dimensionless time by $t_0 q_N^c(N)$, and from it the absolute performance can be found.

A number of absolute performance metrics have been defined in Chapter-2 which are all scaled inverse-time metrics of the form

$$R(N;p) = \frac{F(N)}{T(N;p)} \tag{4.29}$$

where $F(N)$ is a scale factor depending only on the problem size, and is therefore a constant if one is studying the scaling with p of a constant sized problem. These absolute performance measures must be used if one is comparing the performance of one computer with another, as is the case in benchmarking.

If, however, one is studying the scaling of a program on a particular computer (and not comparing with other computers) it is legitimate to factor out the performance of a single processor, and express the absolute performance as a product of the single-processor performance and a scaling function, thus

$$R(N;p) = S_p(N;p)R(N;1) \tag{4.30}$$

The scaling function is, of course, the traditional and familiar self-speedup measure

$$S_p(N;p) = \frac{T(N;1)}{T(N;p)} = \frac{R(N;p)}{R(N;1)} \tag{4.31}$$

where $T(N;1)$ and $R(N;1)$ are the execution time and corresponding performance of the program on one processor of the same computer. For this reason we call the scaling function the self-speedup to distinguish it from other definitions of speedup in which the one-processor time may be taken as the execution time on some other reference computer. Snelling [71] has called this definition of speedup the *relative speedup*.

The self-speedup may also be regarded from the last equality of equation (4.31) as the absolute performance expressed as a dimensionless ratio to the absolute performance on one processor. Speedup is therefore the dimensionless performance and is unaffected by a constant scale factor applied to the time, whether $F(N)$ in equation (4.29) or $q_N^c(N)t_0^c$ in equation (4.25). Thus in terms of dimensionless time

$$S_p(N;p) = \frac{T'(N;1)}{T'(N;p)} \tag{4.32}$$

The one-processor time contains no communication, and comprises only the first term of equation (4.25), hence

$$T'(N;1) = \delta_3 s_p^s(1) \tag{4.33}$$

and

$$T'(N;p) = \delta_3 s_p^s(p) + \delta_2 s_p^c(p) + q_p^c(p) \tag{4.34}$$

Then substituting in equation (4.32) we have, in general,

$$S_p(\delta_3, \delta_2, p) = \frac{\delta_3 s_p^s(1)}{\delta_3 s_p^s(p) + \delta_2 s_p^c(p) + q_p^c(p)} \tag{4.35}$$

and at the optimum, when $p = \tilde{p}$

$$\tilde{S}_p(\delta_3, \delta_2, \tilde{p}(\delta_3, \delta_2)) = \tilde{S}_p(\delta_3, \delta_2) = \frac{\delta_3 s_p^s(1)}{\delta_3 s_p^s(\tilde{p}) + \delta_2 s_p^c(\tilde{p}) + q_p^c(\tilde{p})} \tag{4.36}$$

where we have shown the functional dependencies of the speedup in parentheses.

4.2.3 Dimensionless Universal Scaling Diagram (DUSD)

Equations (4.35) and (4.36) show that the speedup in general depends on the two dimensionless ratios and the number of processors, but that the speedup at the optimum point using $p = \tilde{p}$ depends only on the two dimensionless ratios. This latter fact enables us to draw a Dimensionless Universal Scaling Diagram (DUSD), in which contours of constant optimum speedup $\tilde{S}_p$ and constant optimum number of processors $\tilde{p}$ are drawn on the (δ_2, δ_3) parameter plane. In general, the dimensionality of this 'plane' is one less than the number of hardware parameters used to describe the computer, or one less than the number of terms in the timing formula (4.10). In practice we have found it more useful to use a derived parameter δ_1 instead of δ_2, and to plot in the (δ_1, δ_3) plane, where

$$\delta_1 = \frac{\delta_3}{\delta_2} = \frac{s_N^s(N)}{s_N^c(N)} \left(\frac{r_\infty^c}{r_\infty^s}\right) \tag{4.37}$$

$$\delta_3 = \frac{s_N^s(N)}{q_N^c(N)} \left(\frac{1}{t_0^c r_\infty^s}\right) \tag{4.38}$$

If we use the concept of *computational intensity* [54, 53] which is defined as the ratio of arithmetic work to words transferred, then the δ_1 axis is the computational intensity axis because this axis is proportional to the ratio of the computational intensity of the program (s_N^s/s_N^c) to the *half-performance computational intensity* of the hardware (r_∞^s/r_∞^c). For this reason we call δ_1 the dimensionless computational intensity (or dimensionless intensity for short).

In terms of these two ratios, the speedup can be expressed as follows

$$\tilde{S}_p(\delta_1, \delta_3) = \left[\frac{s_p^s(1)}{s_p^s(\tilde{p})}\right] / \left[1 + \frac{1}{\delta_1}\frac{s_p^c(\tilde{p})}{s_p^s(\tilde{p})} + \frac{1}{\delta_3}\frac{q_p^c(\tilde{p})}{s_p^s(\tilde{p})}\right] \tag{4.39}$$

This way of expressing the self-speedup is revealing because the numerator alone expresses perfect self-speedup with the performance being proportional to the number of processors (s_p^s usually being $1/p$). The deviations of the denominator from unity therefore represent undesirable overheads that degrade performance from the ideal. The first of these is the degradation due to insufficient arithmetic compared to communication (i.e. too small a computational intensity), and is made less important by higher values for the dimensionless intensity δ_1. The last term of the denominator gives the degradation due to message latency, and can be made less important by higher values for the dimensionless work δ_3. Thus higher values for both dimensionless work and intensity, bring the performance closer to the ideal.

Any combinations of the dimensionless parameters can be used to define the axes of the dimensionless plane, but the above choice has the following advantages, and might be adopted as a standard unless there is good reason to do otherwise. Starting from a given inspection point in the diagram, corresponding to given values of $r_\infty^s, r_\infty^c, t_0^c$ and N, and therefore of δ_1 and δ_3, then:

1. In general, higher speedup lies to the top and right of the diagram.

2. The contours of constant speedup should be considered as a performance hill that is to be climbed in the most advantageous way.

3. The y-axis, δ_3, is the message latency axis, because any improvement (i.e. reduction) in latency t_0^c is a movement vertically upward in the diagram.

4. The x-axis, δ_1, is the asymptotic bandwidth axis, because any improvement (i.e. increase) in r_∞^c is a movement horizontally to the right in the diagram.

5. Any improvement in the arithmetic performance, r_∞^s, worsens the speedup, and drives the inspection point diagonally towards the bottom left of the DUSD. The inverse of this is the the well-known phenomenon that it is easy to get good scaling behaviour if

you use slow enough processors, and thereby move diagonally to the upper right of the diagram.

6. A desirable movement diagonally up and to the right can also be achieved by decreasing the latency by the same factor that is used to increase the asymptotic bandwidth. That is to say a movement that keeps the half-performance message length, $n^c_{\frac{1}{2}}$, constant.

7. Lines of constant $n^c_{\frac{1}{2}}$ are diagonals from the lower left to upper right. Higher (i.e. worse) values of $n^c_{\frac{1}{2}}$ lie to the bottom right, and lower values to the top left.

8. The movement of the inspection point as the problem size increases depends on the particular form of the functions $s^s_N(N), s^c_N(N)$ and $q^c_N(N)$. However, in the FFT1 example shown below, it is a motion upward and slightly to the right of the vertical.

After having obtained the optimum self-speedup from the values of δ_1 and δ_3 and the DUSD, or from equations (4.37), (4.38) and (4.39), any of the absolute performance metrics can be obtained by multiplying the self-speedup by the appropriate one-processor performance. For example, the Temporal and Benchmark performances are, respectively

$$\tilde{R}_T(N;\tilde{p}) = \tilde{S}_p/T(N;1) \tag{4.40}$$

$$\tilde{R}_B(N;\tilde{p}) = \tilde{S}_p\, r^s_\infty \tag{4.41}$$

Since the plotted values of self-speedup are dimensionless, all values and axes in the DUSD are dimensionless. The diagram therefore gives the optimum dimensionless performance in terms of the two key dimensionless ratios that determine the parallel performance and scaling. The absolute performance is obtained by multiplying the dimensionless performance by the known single processor absolute performance, r^s_∞. Hence, this single diagram predicts the absolute performance of all programs in the class defined by timing relation (4.25) on all computers describable by the three hardware parameters, and all problem sizes.

All the above analysis carries through if the computer parameters $r^s_\infty(N), r^c_\infty(N)$ and $t^c_0(N)$ are functions of the problem size N, and we will make use of this fact in the FFT1 example below. However the analysis fails if the computer parameters are a function of p.

4.3 Computational Similarity

In analogy to the concept of *Dynamical Similarity* in fluid dynamics [13], and based on the properties of the DUSD, we can now enunciate the principle of *Computational Similarity*:

> If two calculations have the same values of the two dimensionless ratios, then they are said to be computationally similar, and will have the same optimum self-speedup, and the same optimum number of processors.

The dimensionless ratios may be any two of the three deltas defined above.

In fluid dynamics the dimensionless numbers are proportional to the ratio of different forces in the fluid, for example

$$\text{Reynolds}' \text{ Number} = \frac{\text{Inertia}}{\text{Friction}} \tag{4.42}$$

$$\text{Froude's Number} = \frac{\text{Inertia}}{\text{Gravity}} \tag{4.43}$$

Similarly in computation, the dimensionless deltas are proportional to the ratio of different terms in the timing expression

$$\delta_3 = \frac{\text{Arith}}{\text{Startup}} \tag{4.44}$$

$$\delta_2 = \frac{\text{Stream}}{\text{Startup}} \tag{4.45}$$

$$\delta_1 = \frac{\text{Arith}}{\text{Stream}} \tag{4.46}$$

So that computationally similar calculations have similar proportions of time spent on the different activities.

4.4 Application to the Genesis FFT1 Benchmark

As an example of the above procedure, we have analysed the published results on the Intel iPSC/860 for the Genesis FFT1 benchmark (Hey et.al. [35]) using the three parameter timing relation given by Getov [30]:

$$T(N;p) = \frac{5N\log_2 N}{r_\infty^s p}\ \text{flop} + \frac{N\log_2 p}{r_\infty^c p}\ \text{w} + t_0^c \log_2 p \tag{4.47}$$

Comparing with the general formulae derived in the section-4.2.1, we have

$$s^s(N;p) = \frac{5N\log_2 N}{p}\ \text{flop}; \quad s_N^s(N) = 5N\log_2 N\ \text{flop}; \quad s_p^s(p) = \frac{1}{p} \tag{4.48}$$

$$s^c(N;p) = N\frac{\log_2 p}{p}\ \text{w}; \quad s_N^c(N) = N\text{w}; \quad s_p^c(p) = \frac{\log_2 p}{p} \tag{4.49}$$

$$q^c(N;p) = \log_2 p; \quad q_N^c(N) = 1; \quad q_p^c(p) = \log_2 p \tag{4.50}$$

The timing relation is made dimensionless, and converted to natural logarithms, by dividing through by $t_0^c \log_2 e$

$$T'(N;p) = \frac{T(N;p)}{t_0^c \log_2 e} = \delta_3' \frac{1}{p} + \delta_2 \frac{\log_e p}{p} + \log_e p \tag{4.51}$$

where we have introduced a prime to show that natural, rather than base-2, logarithms are used in the following definitions

$$\delta_3' = \frac{5N\log_e N}{t_0^c r_\infty^s}\ \text{flop} \tag{4.52}$$

$$\delta_2 = \frac{N}{t_0^c r_\infty^c}\ \text{w} \tag{4.53}$$

$$\delta_1' = \frac{\delta_3'}{\delta_2} = \frac{5\log_e N}{r_\infty^s / r_\infty^c} \frac{\text{flop}}{\text{w}} \tag{4.54}$$

The results are expressed as a Benchmark performance [47, 48], defined as

$$R_B(N;p) = \frac{F_B}{T(N;p)} \tag{4.55}$$

$$\text{where} \quad F_B(N) = 5N\log_2 N\ \text{flop} \tag{4.56}$$

Substituting equations (4.48), (4.49) and (4.50) into (4.28), the optimum equation for $\tilde{p}$ is

$$\tilde{p} = \delta_3' + \delta_2(\log_e \tilde{p} - 1) \tag{4.57}$$

Equation (4.57) is transcendental and has no explicit solution, however it may be easily solved by iteration. The last value (or initially a guess) for $\tilde{p}$ is put in the right-hand side and the left-hand side delivers a closer value to the solution. About ten iterations are adequate for most purposes.

The dimensionless time, both for general p and at the optimum $p = \tilde{p}$, is then given by:

$$T'(N;1) = \delta_3' \tag{4.58}$$

$$T'(N;p) = \frac{\delta_3'}{p} + \delta_2 \frac{\log_e p}{p} + \log_e p \tag{4.59}$$

$$\tilde{T}'(N;\tilde{p}) = 1 + \frac{\delta_2}{\tilde{p}} + \log_e \tilde{p} \tag{4.60}$$

and the dimensionless performance, or self-speedup is for general p

$$S_p(N;p) = \frac{T'(N;1)}{T'(N;p)} = \frac{\delta_3'}{\frac{\delta_3'}{p} + \delta_2 \frac{\log_e p}{p} + \log_e p} \tag{4.61}$$

and at the optimum performance point, $p = \tilde{p}$

$$\tilde{S}_p(N;\tilde{p}) = \frac{T'(N;1)}{\tilde{T}'(N;\tilde{p})} = \frac{\delta_3'}{1 + \frac{\delta_2}{\tilde{p}} + \log_e \tilde{p}} \tag{4.62}$$

In order to draw the constant speedup curves on the DUSD diagram, we need simultaneously to solve the optimum equation (4.57) and the speedup equation (4.62). Ideally we would like an explicit equation giving δ_3' as a function of δ_2 along the contour. That is to say, to eliminate $\tilde{p}$ between the above two equations. However, the transcendental nature of the equations makes this impossible, but we can use $\tilde{p}$ as a parametric variable (like the length) that changes value along the contour line. Treating equations (4.57) and (4.62) as two linear equations for δ_3' and δ_2, and $\tilde{p}$ and $\tilde{S}$ as constants, we can solve explicitly for the deltas in terms of the speedup and optimum number of processors along the contour:

$$\delta_3' = \frac{\tilde{S}_p \log_e^2 \tilde{p}}{\log_e \tilde{p} - 1 + \frac{\tilde{S}_p}{\tilde{p}}} \tag{4.63}$$

$$\delta_2 = \frac{\tilde{p} - \tilde{S}_p(\log_e \tilde{p} + 1)}{\log_e \tilde{p} - 1 + \frac{\tilde{S}_p}{\tilde{p}}} \tag{4.64}$$

$$\delta_1' = \frac{\delta_3'}{\delta_2} = \frac{\tilde{S}_p \log_e^2 \tilde{p}}{\tilde{p} - \tilde{S}_p(\log_e \tilde{p} + 1)} \tag{4.65}$$

where δ_1' will be used as one of the axes for the DUSD.

Given the above formulae the DUSD can be drawn as follows: For each of a selected number of self-speedup values (or contours values), δ_3' and δ_1' are tabulated for a substantial number of values of $\tilde{p}$ between 1 and 10 000 (the maximum number of processors of current interest), roughly equally spaced logarithmically (about ten values for each decade). For each $\tilde{S}_p$, the values of δ_3' are then plotted against δ_1', to give the contour for that value of optimum self-speedup or dimensionless performance.

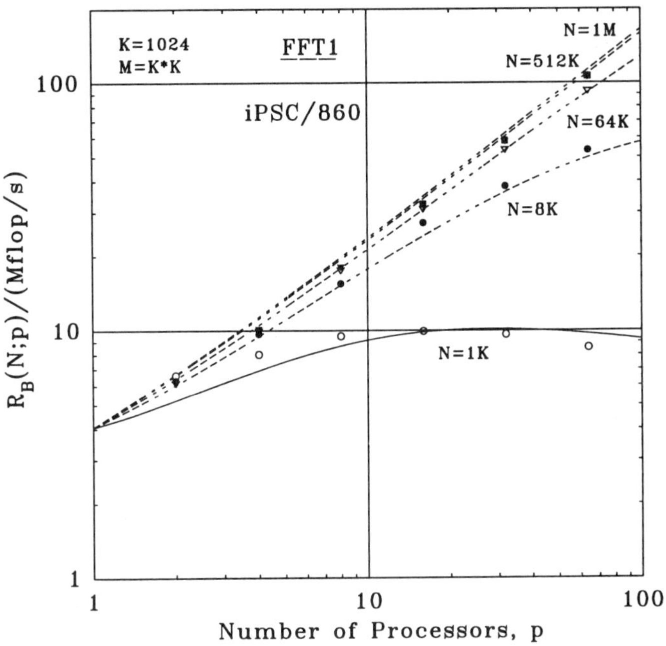

Figure 4.2: A fit to the measured Benchmark performance for the Genesis FFT1 benchmark on the Intel iPSC/860, using three hardware parameters. Symbols are measured values and lines are the theoretical fit using equations (4.47) and (4.55).

4.4.1 Three-paramcter fit to tlie FFT1

The first stage in assessing the validity of the three-parameter timing model (4.47), is to fit the published values of Benchmark performance, $R_B(N;p)$, to equations (4.47) and (4.55). Figure-4.2 shows the result of fitting the 26 measurements of performance to the three parameters

$$r^s_\infty = 4.06 \pm 0.32 \text{ Mflop}_{64}/\text{s} \tag{4.66}$$

$$r^c_\infty = 0.188 \pm 0.018 \text{ Mw}_{64}/\text{s}; \qquad t^c_0 = 762 \pm 49 \ \mu\text{s} \tag{4.67}$$

The square root of the sum of the squares (the norm) of the residuals for this fit was 0.23, and the asymptotic standard deviation for each parameter is given after the $\pm$ symbol. The key hardware combinations are

$$n^{cs}_{\frac{1}{2}} = t^c_0 r^s_\infty = 3090 \text{ flop}_{64}; \qquad n^c_{\frac{1}{2}} = t^c_0 r^c_\infty = 143 \text{ w}_{64} \tag{4.68}$$

$$\left(\frac{r^s_\infty}{r^c_\infty}\right) = 21.6 \frac{\text{flop}_{64}}{\text{w}_{64}} \tag{4.69}$$

The last of these is the second combination of hardware parameters identified by Solchenbach as key to determining the communication performance of a parallel computer [72], and was called by him the '8B-Transfer time in flop'. Fox has also emphasised the importance of this ratio which he calls (tcomm/tcalc). This ratio of asymptotic rates tells how much faster the arithmetic is compared to the communication of data, and has previously been called by the author the 'half-performance computational intensity' [54, 53]. For efficient calculation the computational intensity of the program (arithmetic operations per data word communicated from other processors) must be significantly higher than this ratio.

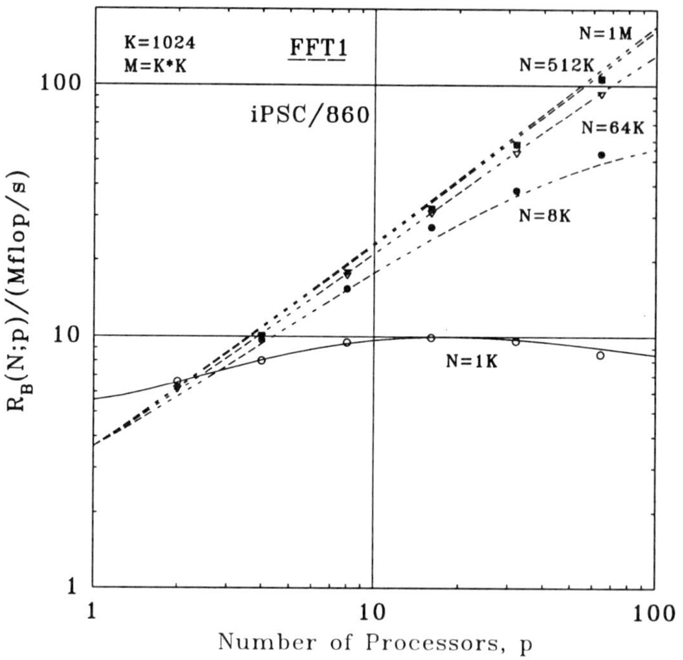

Figure 4.3: An improved fit to the measured Benchmark performance for the Genesis FFT1 benchmark on the Intel iPSC/860, using four hardware parameters. Symbols are measured values and lines are the theoretical fit using equations (4.47) and (4.55).

To these hardware combinations we can add the effect of problem size by computing the key dimensionless 'delta' ratios which determine the scaling. Evaluated for $N = 1\text{K} = 1024$ they are

$$\delta_3' = \frac{5N\log_e N}{t_0^c r_\infty^s} \; \text{flop}_{64} = 11.5 \tag{4.70}$$

$$\delta_2 = \frac{N}{t_0^c r_\infty^c} \; \text{w}_{64} = 7.17 \tag{4.71}$$

$$\delta_1' = \frac{5\log_e N}{r_\infty^s / r_\infty^c} \; \frac{\text{flop}_{64}}{\text{w}_{64}} = 1.60 \tag{4.72}$$

Inspection of Fig-4.2 shows that the three parameter fit is satisfactory (say within 10%) for all problem sizes except $N = 1\text{K}$, in which case the measured performance is twenty to thirty percent higher than is predicted by the timing formula. However, the peak in performance around 20 to 30 processors for $N = 1\text{K}$ is predicted satisfactorily by the three parameter model.

The higher than expected performance for the small problem is thought to be due to the effect of the cache. The data for a 1K FFT when split over two or more processors is likely to fit into the 8KB cache of the Intel i860 chip, hence the average arithmetic rate will be higher than that observed by larger problems that do not fit in the cache. The fit to the experimental data can be made almost exact if we introduce a fourth parameter, γ, namely the ratio of in-cache arithmetic rate r_∞^{s2} to the out-of-cache arithmetic rate r_∞^{s1}. The in-cache rate is applied to $N = 1\text{K}$ and the out-of-cache rate to all the other larger problem sizes. Thus r_∞^s has become a function of problem size, but we have previously noted that the scaling analysis carries through exactly as before. The only difference is that different values of r_∞^s are used to calculate the delta values for the different problem sizes. This alters the point on the DUSD that is used to read off the self-speedup and optimum number of processors. It does not however change the DUSD surface itself and or its contours in Figure-4.4. Figure-4.3 shows the four parameter fit using

$$\gamma = 1.52 \pm 0.13 \quad (4.73)$$

$$r_\infty^{s2} = 5.57 \pm 0.47 \text{ Mflop}_{64}/\text{s}; \quad r_\infty^{s1} = 3.65 \pm 0.22 \text{ Mflop}_{64}/\text{s} \quad (4.74)$$

$$r_\infty^{c} = 0.218 \pm 0.019 \text{ Mw}_{64}/\text{s}; \quad t_0^c = 847 \pm 31\ \mu\text{s} \quad (4.75)$$

In this improved fit, the introduction of the fourth parameter has reduced the norm of the residuals to 0.16.

4.4.2 DUSD for FFT1

The dimensionless universal scaling diagram for the FFT1 benchmark, or any program in the same class defined by a timing relationship of the form of equation (4.51), is shown in Fig-4.4. Both axes are logarithmic in order to encompass as large a range in the dimensionless ratios as possible. The x-axis is δ_1', and is proportional to the asymptotic communication bandwidth and weakly dependent on N through the logarithm. The y-axis is δ_3', and is inversely proportional to the message latency, and proportional to the problem size times its logarithm. Both axes are inversely proportional to the arithmetic performance, r_∞^s. The solid lines are contours of constant optimum self-speedup, $\tilde{S}$, and are equally spaced roughly logarithmically from 1.001 to to 100. The dotted lines are lines of constant optimum number of processors, $\hat{p}$ from 2 to 1000 also roughly equally spaced logarithmically.

The solid line rising almost vertically is the path followed by the FFT1 benchmark when executing on the Intel iPSC/860 as its problem size varies from $N = 100$ to $N = 10^5$. This is calculated using the three-parameter fit given in equation (4.66). Suppose one wishes to solve a problem with $N = 1000$, the DUSD shows that about 25 processors are optimum, and that the resulting self-speedup would be about 2.5. We can now ask what is the best improvement to the communication hardware for this problem size. One option would be to decrease the message latency by a factor ten, which causes a movement vertically upwards in the diagram by a factor ten in the δ_3 axis. The DUSD shows that this would result in an optimum self-speedup of about 18, but would require almost 500 processors. The only way to reduce the required number of processors is to move right in the diagram, which means to increase the communication bandwidth. If this were increased by a factor ten the self-speedup would only be improved marginally to about 20, but this could be achieved with far fewer processors, namely about 150. It seems, therefore, that the best action is to move diagonally upwards and to the right, which means to decrease the latency by the same factor that is used to increase the bandwidth. This means that the product of latency and bandwidth is kept constant. This product is, in fact, $n_{\frac{1}{2}}^c$ the half-performance message length for communication (see equation (4.11)) [45, 46, 52, 50]. In so far as moving towards the top right of the DUSD is desirable, a good general rule seems to be to improve communication hardware in such a way that $n_{\frac{1}{2}}^c$ remains roughly constant or decreases.

In order to predict the performance for another computer, perform the following steps:

1. Measure or estimate the three computer hardware parameters, r_∞^s, r_∞^c and t_0^c .

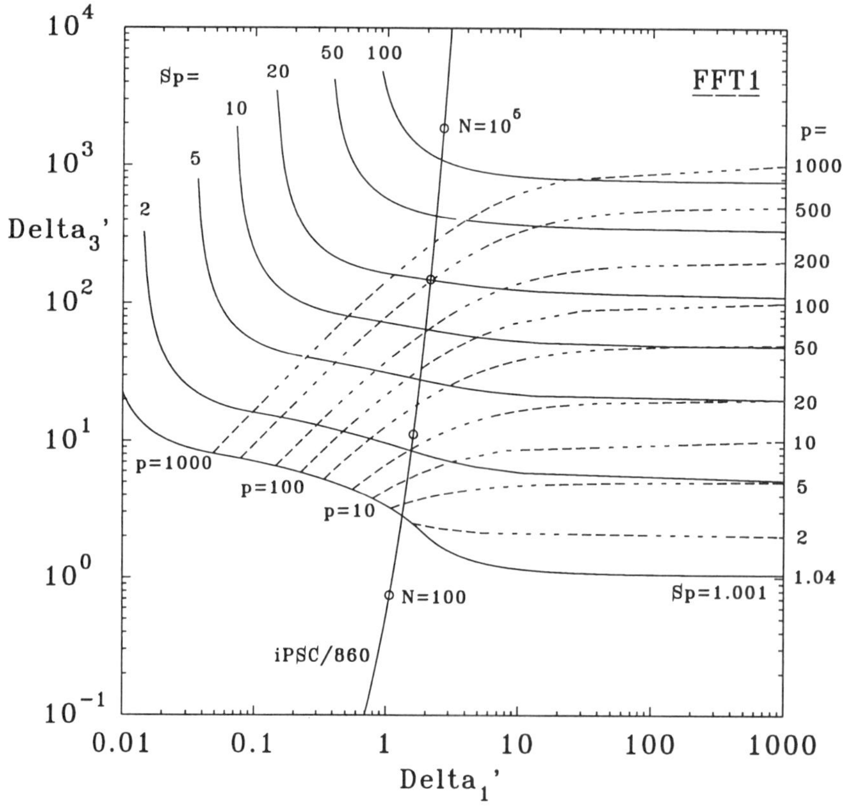

Figure 4.4: The Dimensionless Universal Scaling Diagram (DUSD) for the Genesis FFT1 benchmark and any program with a timing relation of the form of equation (4.51), for all computers representable by the three hardware parameters and for all problem sizes.

2. Calculate the two dimensionless ratios δ_3' and δ_1', from equation (4.70) and (4.72). This introduces and takes into account the effect of problem size N.

3. Read off from the DUSD the optimum number of processors to use, $\tilde{p}$, and the optimum self-speedup, $\tilde{S}_p$. If greater accuracy is required, calculate these quantities from equations (4.57) and (4.62).

4. Multiply the optimum self-speedup by the one-processor performance r_∞^s to obtain the optimum absolute Benchmark performance $\tilde{R}_B$.

The solid contour lines for constant self-speedup which appear above the dotted line for $\tilde{p} = 1000$ have been drawn on the assumption that 1000 is the maximum number of processors available, and this accounts for the change in shape of the contours. If one always uses the maximum number of processors p_{max} then the equation for the contour follows immediately from the general speedup equation (4.61), by rearranging δ_3' as a function of δ_1'

$$\delta_3' = \frac{\tilde{S}_p \log_e p_{max}}{1 - \frac{\tilde{S}_p}{p_{max}}\left(1 + \frac{\log_e p_{max}}{\delta_1'}\right)} \tag{4.76}$$

Chapter 5

Presentation of Results

The usefulness of any benchmark suite is determined, to a large extent, by the availability of both the source of the benchmarks themselves and the results obtained with them. It is therefore desirable that both of these be in the public domain and easily available. In fact the Parkbench committee has made this one of their aims, via the establishment of a public-domain database/repository for benchmark programs and results. With the advent of the Internet and the World-Wide Web, it is now possible to make such a data-base easily available from anywhere in the world, and thus to realise the above aim.

However this requires specialised computer software to setup and maintain the database. The first pioneering work in this direction was taken by Jack Dongarra[24] and his group at the Argonne National Laboratory during the 1980s, when they established a database for the results of the Linpack and other benchmarks and associated literature which was called Netlib. This operated automatically via the Internet e-mail service. An e-mail message sent to a particular destination, and containing a specific request for a benchmark or report was automatically scanned, and the required information (perhaps the latest list of Linpack results) was automatically e-mailed back to the sender. Alternatively the information could be retrieved by the 'anonymous' file-transfer procedure from a continuously running server on the Internet. This can be done by anyone with access to Internet, and the server is able to keep statistics of requests made to it, by examining the e-mail addresses of the users which must be given as the 'password' when loging on to the server. This information service has now been transferred to the University of Tennessee and the Oak Ridge National Laboratory. The latest 'point-and-click' interface to this database, called Xnetlib, uses X-window software and is the subject of section5.1. It can be accessed by any World-Wide Web browser such as, for example, the public-domain Mosaic from NCSA[4, 66].

The design and use of the Xnetlib performance database server (PDS) is described in section-5.2, and a number of example Xnetlib screen displays are shown. This is primarily the Master's thesis work of Brian LaRose and is taken with his permission from Berry, Dongarra, LaRose and Letsche[17] and the Parkbench Report.

The PDS database is built around a relational database, and thereby allows fairly complicated conditional searches to be made of what is becoming a very large database. Such a large database would be quite unusable without this facility, and the result of such a selective enquiry is a table of database entries of manageable size that satisfy the requested condition. Each entry line in the database is about 100 characters and represents a single computer/compiler measurement with only one or a few output numbers. This ideally suits the Linpack benchmark which traditionally gives, for each entry, the performance for two problem sizes ($N = 100$, $N = 1000$) and the theoretical peak performance for a single processor computer/compiler combination.

Parallel benchmarks on the other hand require the storage of performance values for about ten different numbers-of-processors for each of three or even four problem sizes (see e.g. section-2.7). Such a large number ($\approx$ 50) of data points is required for a proper study of the scaling of the parallel benchmark, and to give sufficient information for informed extrapolation. That is to say that one should have enough data to fit a timing/performance model with, say, three to five hardware parameters with statistical significance (see e.g. section-4.4). If the fit is satisfactory, then limited extrapolation can be made with some confidence. The natural way to view and interpret such results is graphically, and the Southampton University Concurrent Computing Group working under the direction of Tony Hey has produced a Graphical Benchmark Information Server (GBIS) which is described in section-5.4 and, like PDS, can be accessed off the World-Wide Web. This is primarily the work of Mark Papiani and is taken with his permission from Papiani, Hey and Hockney [64].

Most of the low-level benchmarks described in Chapter-3 produce data that also require graphical representation. For example, each of the 17 DO-loop kernels of RINF1 produce about 40 measurements of time for different vector or loop lengths, to which a best straight line, described by the (r_∞,$n_{\frac{1}{2}}$) parameters, is fitted by least squares. To assess the quality of the data, and to see the validity (or not) of the straight-line approximation, it is always desirable to examine a scatter plot of 't versus n', and compare it with the straight-line approximation. The same applies to the results of COMMS1, some examples of which are shown in Fig.3.9. The Southampton GBIS allows the detailed results of these benchmarks to be displayed easily in this way.

5.1 Xnetlib

The Netlib software distribution system maintained at the University of Tennessee and Oak Ridge National Laboratory has been in development for several years and has established a large database of scientific (numerical) software and literature. Netlib has over 120 different software libraries, as well as hundreds of articles and bibliographic information.

Originally, Netlib software library access involved the use of electronic mail to form and process queries. However, in 1991, an X-windows interface, *Xnetlib*, was developed by Dongarra et al. [23] in order to provide a more immediate and efficient access to the Netlib software libraries. To date (1994), there have been over 3 200 requests for the Xnetlib tool. In turn, the number of Netlib acquisitions has also escalated. In fact, there were over 86 000 Xnetlib-based transactions and over 150 000 electronic mail acquisitions from Netlib in 1992 alone.

5.2 PDS: Performance Database Server

The process of gathering, archiving, and distributing computer benchmark data is a cumbersome task usually performed by computer users and vendors with little coordination. Within Xnetlib [23] there is a mechanism to provide Internet-access to a performance database server (PDS) which can be used to extract current benchmark data and literature. PDS [56] provides an on-line catalog of public-domain computer benchmarks such as the Linpack Benchmark [24], Perfect Benchmarks [16], and the NAS Parallel Benchmarks [11]. PDS does not reformat or present the benchmark data in any way that conflicts with the original methodology of any particular benchmark; it is thereby devoid of any subjective interpretations of machine performance. PDS is providing a more manageable approach to the development and support of a large dynamic database of published performance metrics.

The PDS system was developed at the University of Tennessee and Oak Ridge National Laboratory and is an initial attempt at performance data management. This on-line database of computer benchmarks is specifically designed to provide easy maintenance, data security, and data integrity in the benchmark information contained in a *dynamic* performance database.

PDS was designed with a simple tabular format that involves displaying the data in rows (machine configuration) and columns (numbers). Graphical representations of tabular data, such as the representation by SPEC [74] with the obsolescent *SPECmarks*, are straightforward.

5.2.1 Design of a Performance Database

Because of the complexity and volume of the data involved in a performance database, it is natural to exploit a database management system (DBMS) to archive and retrieve benchmark data. A DBMS will help not only in managing the data, but also in assuring that the various benchmarks are presented in some reasonable format for users: table or spreadsheet where machines are *rows* and benchmarks are *columns*.

Of major concern is the organisation of the data. It seems logical to organise data in the DBMS according to the benchmarks themselves: a Linpack table, a Perfect table, etc. It would be nearly impossible to *force* these very different presentation formats to conform to a single presentation standard just for the sake of reporting. Individual tables preserve the display characteristics of each benchmark, but the DBMS should allow users to query all tables for various machines. Parsing benchmark data into these tables is straightforward provided a customised parser is available for each benchmark set. In the parsing process, constructing a raw data file and building a standard format ASCII file eases the incorporation of the data into the database.

The functionality required by PDS is not very different from that of a standard database application. The difference lies in the user interface. Financial databases, for example, typically involve specific queries like

EXTRACT ROW ACCT_NO = R103049

in which data points are usually discrete and the user is very familiar with the data. The user, in this case, knows exactly what account number to extract, and the format of retrieved data in response to queries. With our performance database, however, we would expect the contrary: the user does not really know (*i*) what kind of data is available, (*ii*) how to request/extract the data, and (*iii*) what form to expect the returned data to be in. These assumptions are based on the current lack of coordination in (public-domain) benchmark management. The number of benchmarks in use continues to rise with no standard format for presenting them. The number of performance-literate users is increasing, but not at a rate sufficient to expect proper queries from the performance database. Quite often, users just wish to see the best-performing machines for a particular benchmark. Hence, a simple rank-ordering of the *rows* of machines according to a specific benchmark *column* may be sufficient for a general user.

Finally, the features of the PDS user interface should include:

(1) the ability to extract specific machine and benchmark combinations that are of interest,

(2) the ability to search on multiple keywords across the entire dataset, and

(3) the ability to view cross-referenced papers and bibliographic information about the benchmark itself.

We include (3) in the list above to address the concern of proliferating numbers without any benchmark methodology information. PDS would provide abstracts and complete papers related to benchmarks and thereby provide a needed educational resource without risking improper interpretation of retrieved benchmark data.

5.3 PDS Implementation

In this section, we described the PDS tool developed and maintained at the University of Tennessee and the Oak Ridge National Laboratory. Specific topics include the choice of DBMS, the client-server design, and the interface. Specific features of PDS such as `Browse`, `Search`, and `Rank-Ordering` are also illustrated.

5.3.1 Choice of DBMS

Benchmark data is represented by Hobbe's RDB format [38]. This database query language offers several advantages. It is easy to manipulate. It is also efficient, being based on a *PERL* model [77] that uses Unix* pipes to run entirely in memory. We have easily converted raw performance data into RDB format using only a few PERL commands. Additionally, RDB provides several report features that help standardise the presentation of performance data.

The RDB format specifies that database tables are defined using a schema of the form:

01	Computer	35
02	OS/Compiler	45
03	N=100	7N
04	N=1000	7N
05	Peak	7N.

The first column is the field number, the second is the field label, and the third is the field type, in (type-size) format. The default form is ASCII, and N denotes numeric data so that the *Peak* entry, for example, is a size 7 numeric field.

The contents of the data files are expected to be in some regular grammar, usually a space separated columnar format. A PERL script takes a description of the columns and builds a tab-separated file. The schema is converted to a tab-separated file using the RDB command *headchg*. The data files are appended to the end of the schema file, and the resulting flat tab-separated file becomes the rdb format table. RDB uses the schema in the header to process the file. An example from the linpack.rdb is provided below.

```
Computer          OS/Compiler       N=100    N=1000   Peak
35       45       7N         7N     7N
CRAY Y-MP C90 (16 proc. 4.2 ns) CF77 5.0 -Zp -Wd-e68    479    9715    15238
CRAY Y-MP C90 (8 proc. 4.2 ns)  CF77 5.0 -Zp -Wd-e68    468    5994    7619
CRAY Y-MP C90 (4 proc. 4.2 ns)  CF77 5.0 -Zp -Wd-e68    388    3272    3810
CRAY Y-MP C90 (2 proc. 4.2 ns)  CF77 5.0 -Zp -Wd-e68    387    1709    1905
```

This rdb table may then be searched for query matching. An example query for a Linpack Benchmark with `N=100 number equal to 388` is

```
cat linpack.rdb | row N=100 eq 388 | ptbl
```

*Unix is a trademark of AT&T Bell Laboratories.

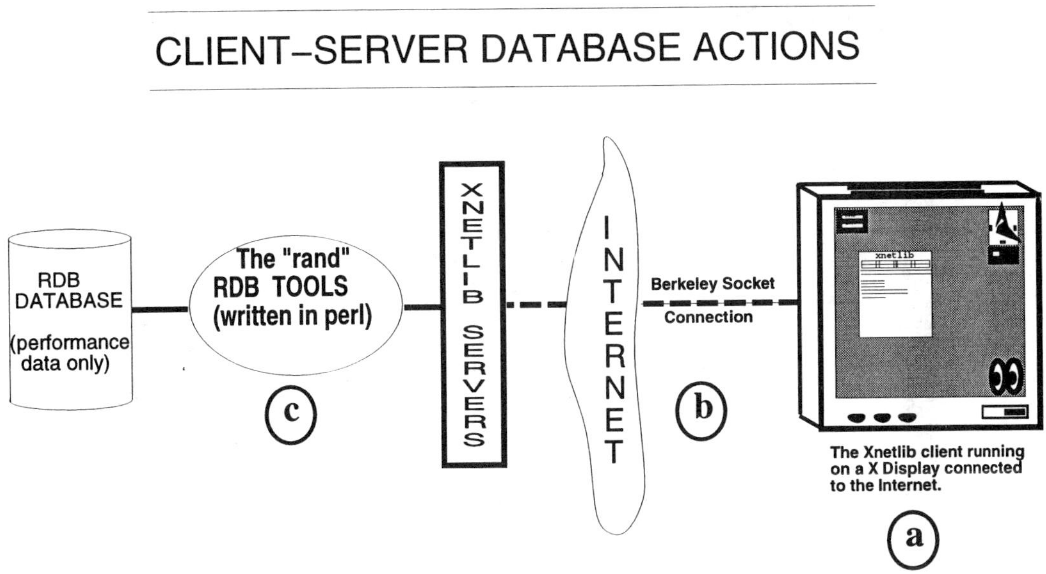

Figure 5.1: The PDS client-server interface: the X workstation (a) communicates over the Internet via Berkeley socket connection (b) to the Xnetlib server, which queries the database using rdb tools (c) and returns benchmark data via the socket connection.

and the benchmark returned is

Computer	OS/Compiler	N=100	N=1000	Peak
CRAY Y-MP C90 (4 proc. 4.2 ns)	CF77 5.0 -Zp -Wd-e68	388	3272	3810

5.3.2 Client-Server Design

Within PDS, the database manager runs only on the server, and the clients communicate via Berkeley sockets to attach to the server and access the database. This functionality was provided by the pre-existing Xnetlib tool and was extended to provide support for the performance data. Figure 5.1 illustrates this client-server interface.

The Xnetlib client is an X-Windows interface that retrieves data from the server. This client is a *view-only* tool which provides the user a *window* into the database yet prohibits data modifications. The `Performance` button under the main xnetlib menu will provide users access to the PDS client tool. Users familiar with Xnetlib 3.0 will have an easy transition to using the PDS performance client.

5.3.3 PDS Features

PDS provides the following retrieval-based functions for the user:

(1) a *browse* feature to allow casual viewing and point-and-click navigation through the database,

(2) a *search* feature to permit multiple keyword searches with Boolean conditions,

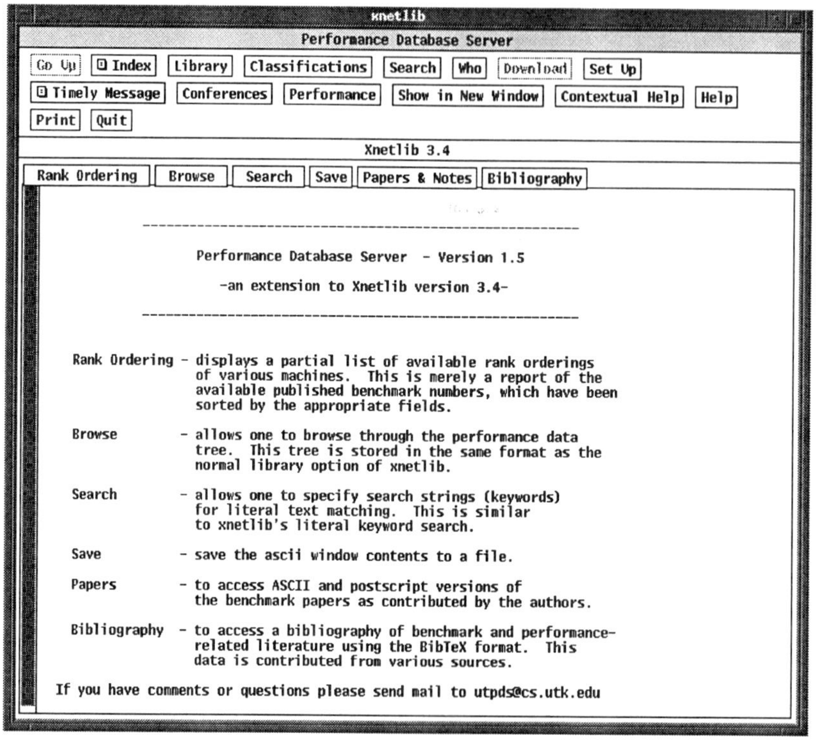

Figure 5.2: PDS start-up window describing available services

(3) a *rank-ordering* feature to sort and display the results for the user, and

(4) a few additional features that aid the user in acquiring benchmark documentation and references.

Figure 5.2 illustrates the PDS start-up window which appears after the user selects the `Performance` button from the Xnetlib 3.3 main menu. The user then selects from the six available services (`Rank Ordering, Browse, Search, Save, Papers & Notes, Bibliography`).

As denoted in Figure 5.3, the `Rank Ordering` option in PDS allows the user to view a listing of machines that have been ranked by a particular performance metric such as megaflops or elapsed CPU time. Both `Rank Ordering` and `Papers & Notes` options are menu-driven data access paths within PDS.

As discussed in [56], the `Rank Ordering` option in PDS allows the user to view a listing of machines that have been ranked by a particular performance metric such as megaflop/s or elapsed CPU time. Both `Rank Ordering` and `Papers` options are menu-driven data access paths within PDS. With the `Browse` facility in PDS, the user first selects the vendor(s) and benchmark(s) of interest, then selects the large `Process` button to query the performance database. The PDS client then opens a socket connection to the server and, using the query language (rdb), remotely queries the database. The `Search` option in PDS permits user-specified keyword searches over the entire performance database. `Search` utilises literal case-insensitive matching along with a moderate amount of aliasing. Multiple keywords are permitted, and a Boolean flag is provided for more complicated searches. Using `Search`, the user has the option of entering vendor names, machine aliases, benchmark names, or specific

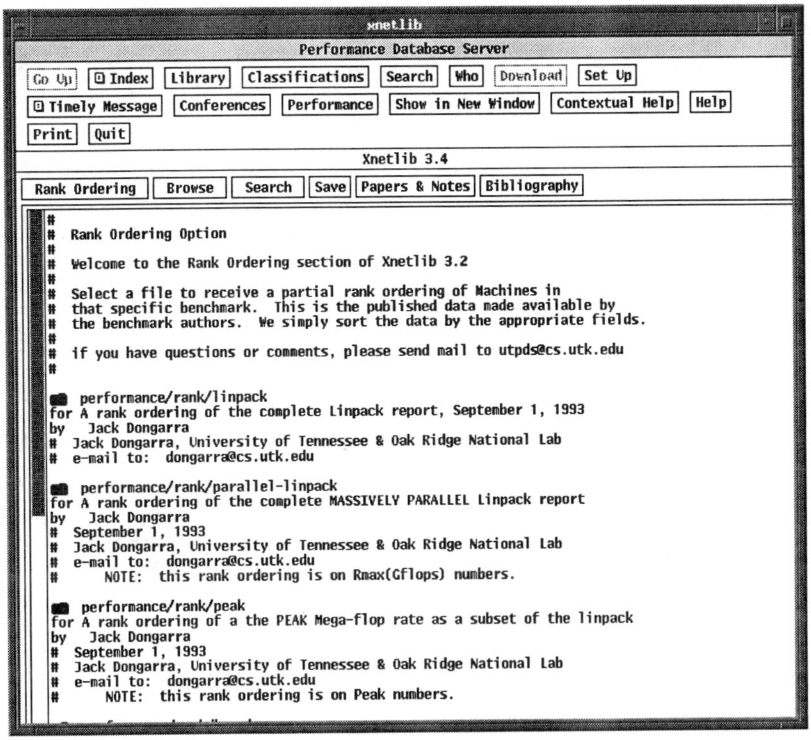

Figure 5.3: The PDS rank-ordering feature menu

strings, or producing a more complicated Boolean keyword search. Since any retrieved data will be displayed to the screen (by default), the **Save** option allows the user to store any retrieved performance data to an ASCII file. Finally, the **Bibliography** option in PDS provides a list of relevant manuscripts and other information about the benchmarks. Future enhancements to PDS include the use of more sophisticated two-dimensional graphical displays for machine comparisons. Additional serial and parallel benchmarks will be added to the database as formal procedures for data acquisition are determined. The **Browse** and **Search** facilities available in the current version of PDS are illustrated in the next section.

5.3.4 Sample Xnetlib/PDS Screens

With the **Browse** facility in PDS (see Figure 5.4), the user first selects the vendor(s) and benchmark(s) of interest, then selects the large **Process** button to query the performance database. The PDS client then opens a socket connection to the server and, using the query language (rdb), remotely queries the database. The format of the returned result is shown in Figure 5.5. Notice that the column headings which will vary with each benchmark. The returned data is displayed as an ASCII widget with scrollbars when needed.

The **Search** option in PDS is illustrated in Figures 5.6 and 5.7. This feature permits user-specified keyword searches over the entire performance database. **Search** utilises literal case-insensitive matching along with a moderate amount of aliasing. Multiple keywords are permitted, and a Boolean flag is provided for more complicated searches. Notice the selection of the Boolean **And** option in Figure 5.6. Using **Search**, the user has the option of entering vendor names, machine aliases, benchmark names, or specific strings, or producing a more

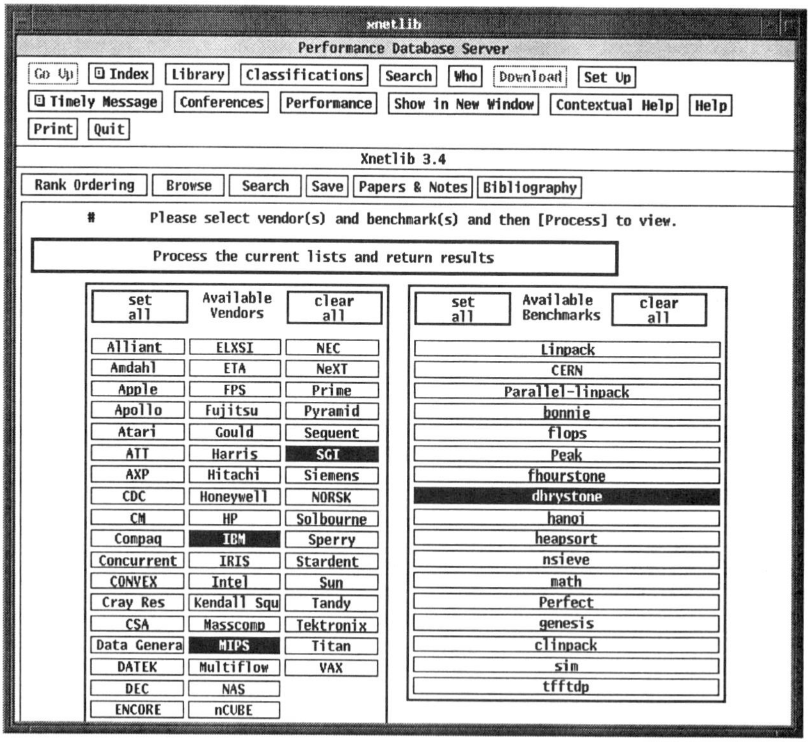

Figure 5.4: The browse facility provided by PDS

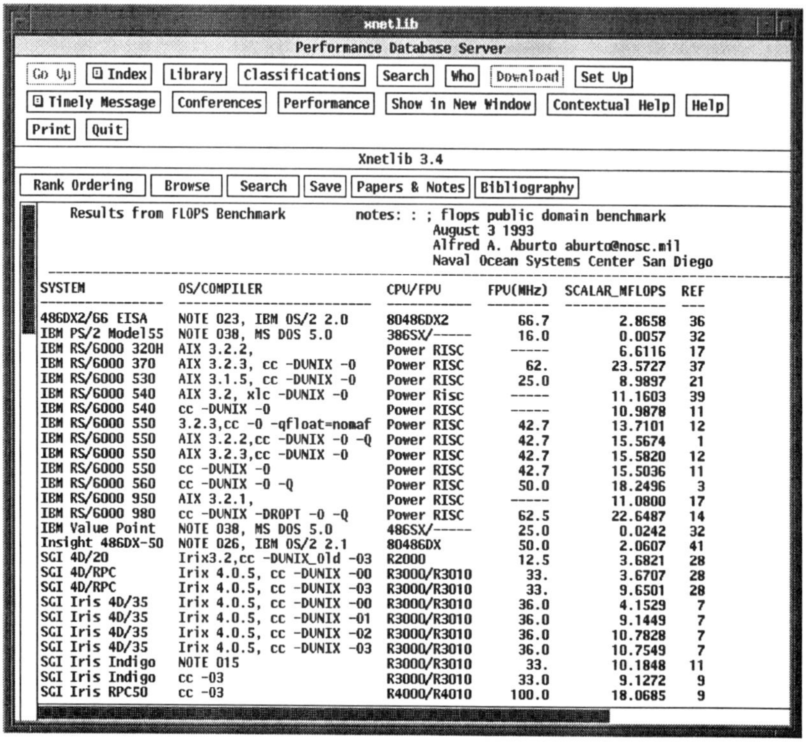

Figure 5.5: Sample data returned by the PDS `Browse` facility

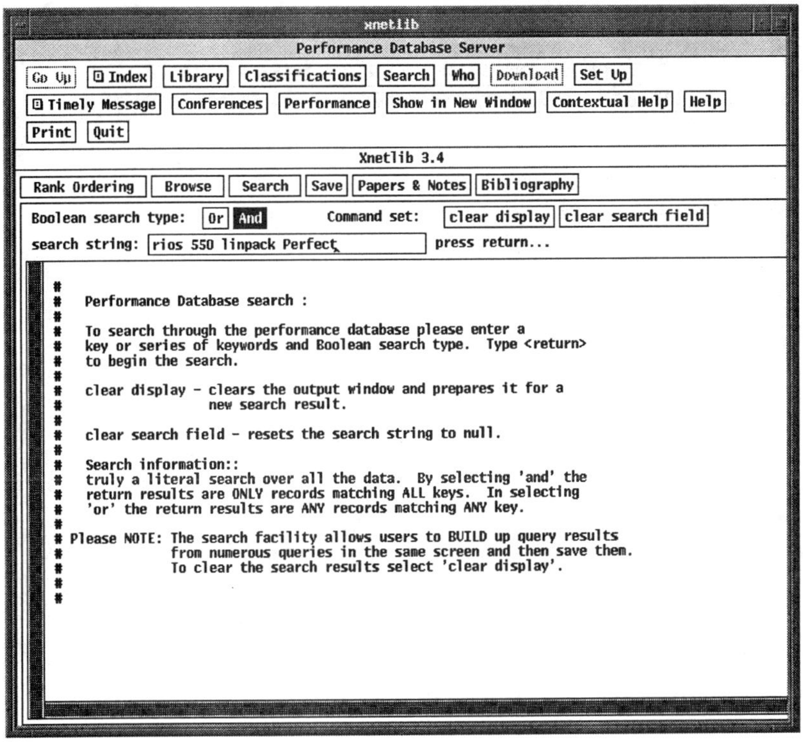

Figure 5.6: Specifying a keyword search using the PDS `Search` facility

complicated Boolean keyword search. The benchmarks returned from the Boolean **And** search

rios 550 linpack Perfect

are shown in Figure 5.7. The alias terms *rios 550* are associated with the IBM RS/6000 Model 550 series of workstations. The specification of *linpack* and *Perfect* will limit the search to the Linpack and Perfect benchmarks only. Since any retrieved data will be displayed to the screen (by default), the `Save` option allows the user to store any retrieved performance data in an ASCII file.

5.3.5 PDS Availability

To receive Xnetlib with PDS support for Unix-based machines, send the electronic mail message *send xnetlib.shar from xnetlib* to `netlib@ornl.gov`. You can *unshar* the file and compile it by answering the user-prompted questions upon installation. Use of *shar* will install the full functionality of Xnetlib along with the latest PDS client tool. Questions concerning PDS should be sent to `utpds@cs.utk.edu`. The University of Tennessee and Oak Ridge National Laboratory will be responsible for gathering and archiving additional (published) benchmark data.

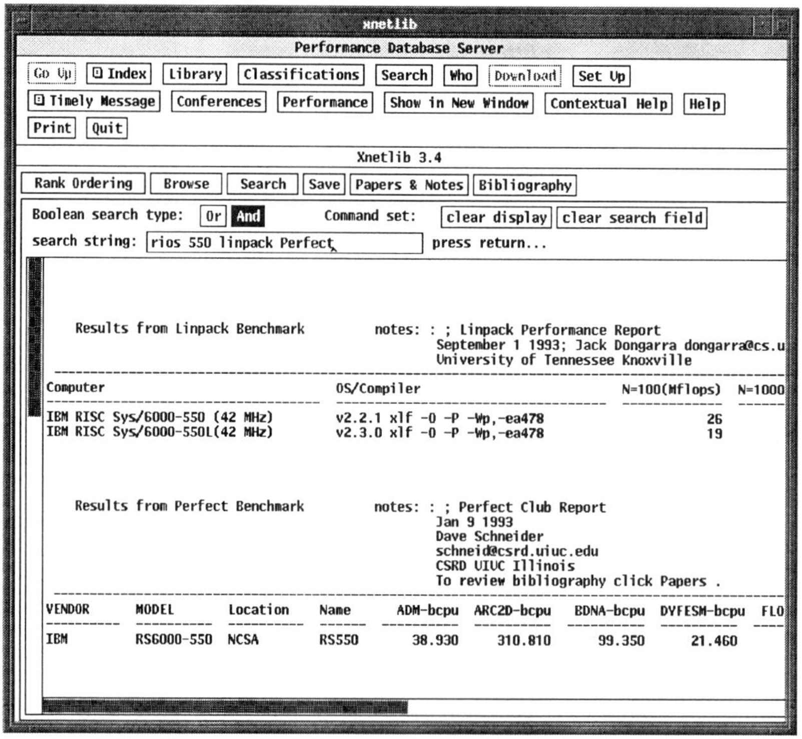

Figure 5.7: Results of a keyword search using the PDS `Search` facility

5.4 GBIS: Interactive Graphical Interface

5.4.1 General Design Considerations

Under PDS each benchmark measurement for a particular problem size N and processor number p, is represented by one line in the PDS database with variable length fields chosen by the benchmark writer as suitable and comprehensive to describe the conditions of the benchmark run. The fields separated by a marker include, benchmarkers name and e-mail, computer location and date, hardware specification, compiler date and optimisation level, N, p, $T(N,p)$, $R_B(N,P)$ and other metrics as deemed appropriate by the benchmark writer. Ideally, the entry line for the database is produced automatically as output by the benchmark program itself. Such a single-line entry is quite satisfactory for single-processor benchmarks which produce a single or small number of results, as has been demonstrated by the Linpack results in the PDS database which has been described above. Performance comparisons between computers can satisfactorily be made by choosing one metric and generating a league table of results using the search facilities of PDS.

Multi-processor benchmarks on highly parallel computers, on the other hand, can yield several tens of numbers for each benchmark on each computer, as factors such as the number of processors and problem size are varied. A graphical display of performance surfaces therefore provides the most satisfactory way of comparing results. This is particularly true when studying the scaling behaviour of performance as the number of processors is increased, which can easily be seen as a graph but is difficult to discern from a table of numbers. To make this possible, the basic data held in the Performance Data Base should be values of $T(N;p)$ for at least 4 values of problem size N, each for sufficient p-values (say 5 to 10) to determine

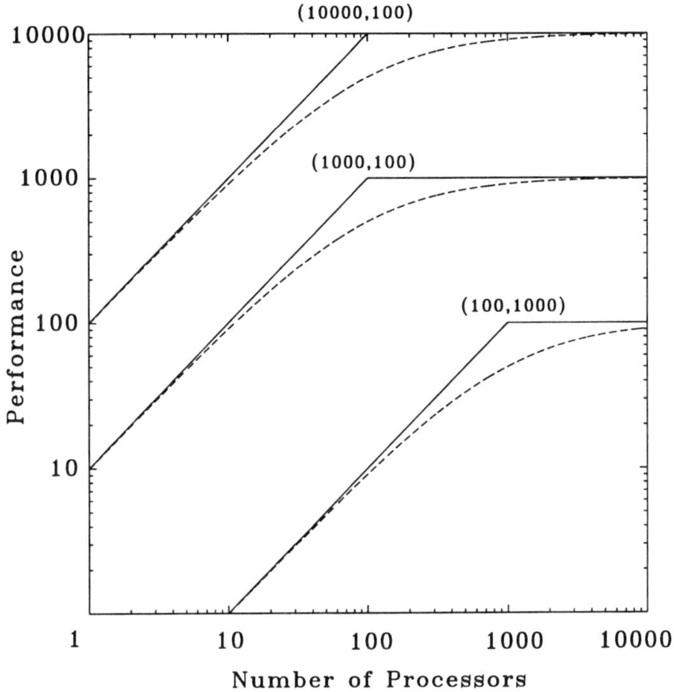

Figure 5.8: Graph showing the invariance in shape of the Amdahl curve (Eqn.(5.1) and dotted line) to its position on a log/log graph. The asymptotes are shown as solid lines, and pass through the point $(R_\infty, p_{\frac{1}{2}})$.

the trend of variation of performance with number of processors for constant problem size. It is important that there be enough p-values to see Amdahl saturation, if present, or any peak in performance followed by degradation. A graphical interface is really essential to allow this multi-dimensional data to be viewed in any of the metrics defined above, as chosen interactively by the user. The user could also be offered (by suitable interpolation) a display of the results in various scaled metrics, in which the problem size is expanded with the number of processors.

In order to encompass as wide a range of performance and number of processors as possible, a log-scale on both axes is unavoidable, and the format and scale range should be kept fixed as far as possible to enable easy comparison between graphs. A three-cycle by three-cycle log/log graph with range 1 to 1000 in both p and Mflop/s would cover most needs in the immediate future. Examples of such graphs are to be found in [47, 1, 2, 35] and section-2.7.

A log/log graph is also desirable because the size and shape of the Amdahl saturation curve is the same wherever it is plotted on such a graph. That is to say there is a universal Amdahl curve that is invariant to its position on any log/log graph. Amdahl saturation is a two-parameter description of any of the performance metrics, R, as a function of p for fixed N, which can be expressed by

$$R = \frac{R_\infty}{(1 + p_{\frac{1}{2}}/p)} = R_\infty \text{pipe}(p/p_{\frac{1}{2}}) \tag{5.1}$$

where R_∞ is the saturation performance approached as $p \to \infty$ and $p_{\frac{1}{2}}$ is the number of processors required to reach half the saturation performance. In general both R_∞ and $p_{\frac{1}{2}}$ are functions of the problem size N. The graphical interface should allow this universal Amdahl curve to be moved around the graphical display, and be matched against the performance data. The changing values of the two parameters (R_∞,$p_{\frac{1}{2}}$) should be displayed as the Amdahl curve is moved. Figure-5.8 shows as dotted lines three Amdahl curves for three different pairs of (R_∞,$p_{\frac{1}{2}}$). The diagonal solid lines at 45 degrees are the three asymptotes for small numbers of processors ($R \propto p$), and the solid horizontal lines are the asymptotes for large numbers of processors ($R = R_\infty$). The two asymptotes meet at the point $(p_{\frac{1}{2}}, R_\infty)$. Simon and Strohmaier [70] have shown that NAS parallel benchmark results can be fitted very well by such Amdahl curves.

As more experience is gained with performance analysis, that is to say the fitting of performance data to parametrised formulae, it is to be expected that the graphical interface will allow more complicated formulae to be compared with the experimental data, perhaps allowing 3 to 5 parameters in the theoretical formula. But, as yet, we do not know what these parametrised formula should must usefully be.

5.5 The Southampton GBIS

As a major step towards realising the above goals the Southampton University Benchmarking Group has produced an interactive graphical front end to the Parkbench database of performance results[64]. Called the Graphical Benchmark Information Service or GBIS, it is available on the World-Wide Web (WWW) to display interactively graphs of user selected benchmark results from the Parkbench and Genesis benchmark suites. The Graphical Benchmark Information Service provides the following features:

1. Interactive selection of benchmark and computers from which a performance graph is generated and displayed.

2. Default graph options that can be changed:

    (a) Available output formats; gif, xbm, postscript or tabular results.

    (b) Selectable ranges or autoscaling.

    (c) Choice of log or linear axes.

    (d) Key to traces can be positioned as required.

3. Links to Parkbench , Genesis and NAS information available on the WWW (including documentation and source codes).

4. Details of how to send in results for inclusion in the Graphical Interface.

The motivation for the GBIS stemmed from the existence of the Genesis [36, 1, 2] and Parkbench [39] distributed-memory benchmark suites.

These benchmark suites contain several categories of benchmark codes which test different aspects of computer performance. The low-level codes will be frozen in the near future, to be replaced by the Parkbench suite which will be used for the GBIS.

The Parkbench suite contains single-processor codes and multi-processor codes which use the PVM (and in the future MPI) message passing interface. Both types of codes produce tens of numbers as results and the most satisfactory way of displaying these numbers is in a graphical manner.

Many of the multi-processor codes are designed to be run numerous times, on the same computer, with the number of processors varied each time. A performance figure in Mflop/s is obtained for each run. By plotting a graph, called a trace, of performance against number of processors, the observed scaleup can be seen i.e. the change in performance which results when more processors are used to solve the same sized problem. By incorporating performance traces for different computers on the same graph, it is possible to compare the relative performance of different computers, as well as the actual scaleup.

Another factor which can be varied for several of the benchmark codes, is the problem size. Traces of temporal performance in e.g. timestep per second against number of processors can be plotted for the same computer, with a different problem size associated with each trace. The observed scaleup can be seen from the relative position of the traces, i.e. the increase in the size of problem that can be solved in the same time period when more processors are used.

The GBIS allows several other types of graphs to be plotted depending on the type of performance metrics that a particular benchmark produces. These performance metrics include Benchmark performance, Temporal performance, Simulation performance, Speedup and Efficiency. These metrics are defined in Chapter-2 and the Parkbench report.

The low-level benchmarks in the Parkbench suite measure performance parameters that characterise the basic machine architecture. These parameters can be used in performance formulae that predict the timing and performance of more complex codes. Chapter-3 describes the low-level parameters. These low-level benchmarks output vectors of numbers which can only be satisfactorily displayed and analysed as graphs. For example, the GBIS allows a choice of two graph types to be plotted for the COMMS1 benchmark from the Parkbench (and Genesis) suites: either message transfer time or transfer rate can be plotted against message length.

5.5.1 GBIS on the WWW

The aim of the GBIS is to make the graphical results for Genesis and Parkbench as widely available as possible. For this reason the method selected to implement the service was to use the World-Wide Web (WWW), and to ensure that any software used was freely available in the public domain. The GBIS was therefore developed using the NCSA (National Center for Supercomputing Applications, University of Illinois) Mosaic Internet information browser (version 2.4 for the X Window System). In particular, the Mosaic Fill-out forms facility was used to create the user interface. Mosaic forms allow user interaction with the WWW beyond the use of hypertext links available within HTML (HyperText Markup Language). It is possible to use widgets on a WWW page including, radio buttons, text entry fields and selection buttons. These provide an ideal method for allowing the user to select results of interest and also to select the desired graph options. The method used by the GBIS to process the forms is to use Unix Bourne Shell scripts running on the WWW server machine, at the University of Southampton.

The user first selects a benchmark. A Bourne Shell script processes this selection and returns a Mosaic form consisting of a list of manufacturers, for which results exist. A number of manufacturers is then selected which produces another Mosaic form displaying all machine results for the chosen benchmark plus manufacturers. Transparent to the user, each machine result corresponds to a data file of cartesian coordinates stored on the WWW server machine. These data files are used in conjunction with the public domain plotting package, GNUPLOT (Unix version 3.5, patchlevel 3.50.1.17, 27 Aug 93) in order to plot the graphs.

The GBIS uses the information processed from the Mosaic forms, to create an appropriate GNUPLOT batch file. This file contains a list of commands which provide a means of setting graph options (e.g. axes scaling, axes types, etc.) to default values or user selected values and for naming the data files to be plotted. The shell scripts then execute GNUPLOT using this batch file, on the WWW server machine, to produce a postscript file of the graph as output.

Depending on the format the user selected for the results, this postscript file is made available for viewing by a hypertext link from a WWW page, or, is converted to another format (e.g. GIF) which can be displayed directly on a WWW page.

The processes described above generate a number of temporary files on the WWW server machine. These files are named using the process id of the Unix shell script which creates them. In this way it is possible for multiple users to simultaneously access the GBIS without any conflict.

The GBIS operation described above is shown in Fig. 5.9, from the user and system points of view.

5.5.2 Directory Structure for Results Files

The results data files associated with each machine are stored as a hierarchy of Unix text files on the WWW server machine. This method was chosen due to the diversity of different vectors of results which need to be stored for the different benchmarks. The convention used for the directory structure is:

benchmark name/machine manufacturer/result file name

where, result file name is in the form:

```
(machine  model)_(date benchmarked DD.MMM.YYY)_
(problem size, if applicable)_(benchmark metric).
```

For example:

```
MULTIGRID/IBM/SP-2_10.AUG.94_CLASSB_PERF.
```

The machine model, date benchmarked and problem size, from the result file name, are used to label traces on the graph; while the benchmark metric is used in choosing labels for the graph axes and in selecting the choice of log or linear axes for the default graph format. For some benchmarks the result of a single run (on a single machine) corresponds to several results files in the GBIS database. This is to allow the results to be plotted in different ways. For example, if the benchmark produces results which can be plotted as any one of, Benchmark performance, Temporal performance or Simulation performance, against the number of processors, then three different files of coordinates must be stored. These are distinguished by using a different benchmark metric in the filename, so that the appropriate file can be found, for the type of graph requested.

5.5.3 Format of Results Files

The actual contents of results data files are in a format that GNUPLOT can use. This consists of lines which contain coordinates and comment lines which begin with a # character. The availability of comment lines was used to incorporate additional information used by the GBIS system. This is explained with reference to the example file contents shown in Fig. 5.15.

The path and filename for this file is;

```
COMMS1/MEIKO/CS-2_26.MAY.94_TIME.
```

The file contents are plotted if the user selects transfer time against message length. For the COMMS1 benchmark, the transfer rate can also be plotted against message length. Therefore, a file, containing different coordinate values, also exists, with a name ending in `_RATE`.

USER ACTIONS	GBIS SYSTEM OPERATION
	1. Display GBIS home page.
2. Select the link to Graphical Benchmark Results (Fig. 5.10).	
	3. Display Mosaic form giving available benchmarks.
4. Select benchmark (Fig. 5.11).	
	5. Obtain HTML file which lists all available manufacturers for the selected benchmark and create Mosaic form (see section-5.5.4).
6. Select manufacturers (Fig. 5.12).	
	7. Obtain one HTML file for each manufacturer selected for the chosen benchmark. Create Mosaic form of available machine results (see section-5.5.4).
8. Select machines, default graph options, or elect to change defaults (Fig. 5.13).	
	9. Store names of results data files, corresponding to each machine selected, in a temporary file.
	10. If default graph options are being used, create temporary GNUPLOT batch file. Run GNUPLOT to create a temporary postscript file, convert to a GIF file and return this as a WWW page. This ends the process if default graph options are accepted.
	11. If change defaults was selected display the Mosaic form showing available graph options.
12. Select graph options (Fig. 5.14).	
	13. If text results are selected display contents of each file, the names of which are listed in the temporary file of results data file names (step 9, above).
	14. Otherwise, create GNUPLOT batch file using data file names from the temporary file, and add GNUPLOT commands to change the default graph options, as requested. Run GNUPLOT to create a temporary postscript file. Provide WWW link to this file if postscript output format was requested or convert postscript file to other requested format and return result as a WWW page.

Figure 5.9: Summary of steps involved in the GBIS operation, from the user and system points of view.

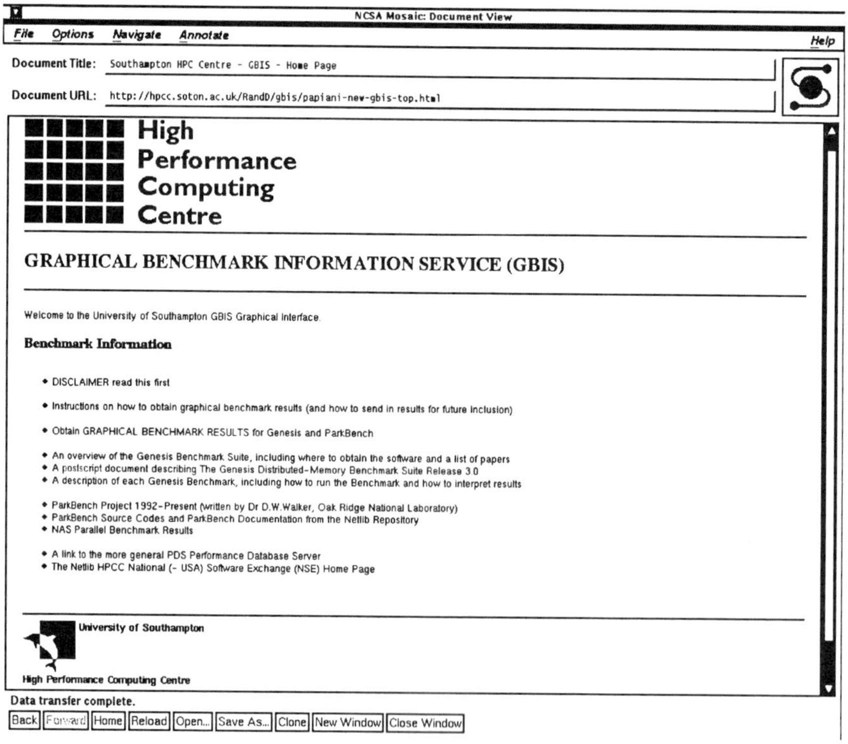

Figure 5.10: GBIS WWW home page.

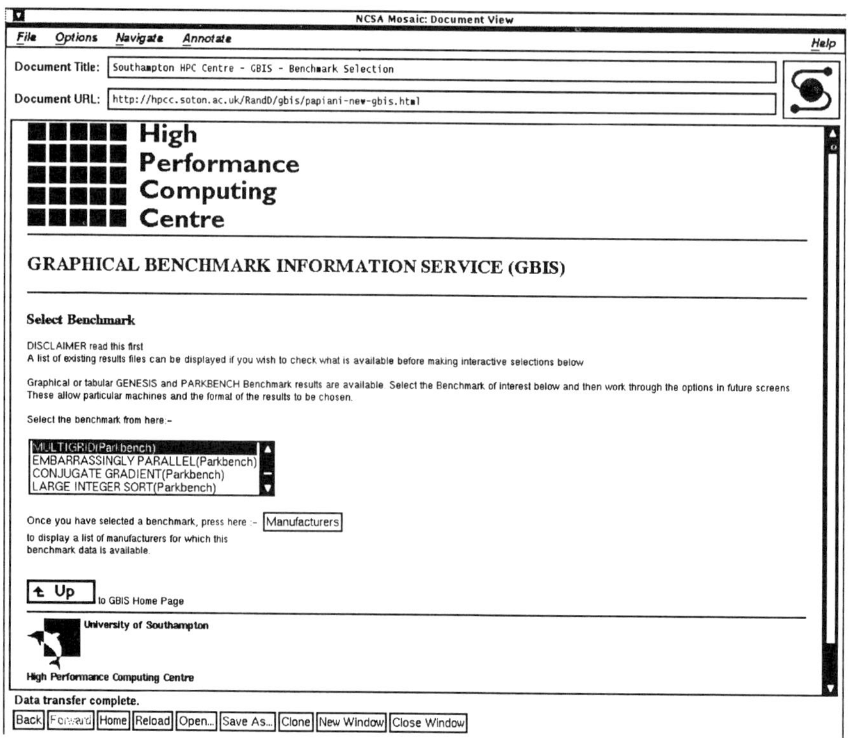

Figure 5.11: GBIS WWW select benchmark page.

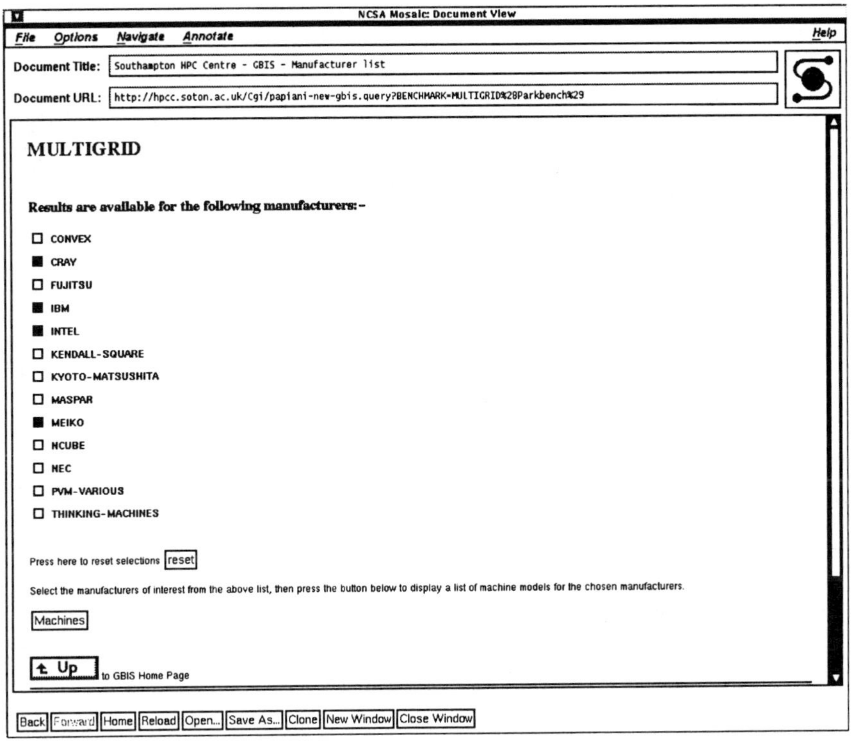

Figure 5.12: GBIS WWW manufacturer list page.

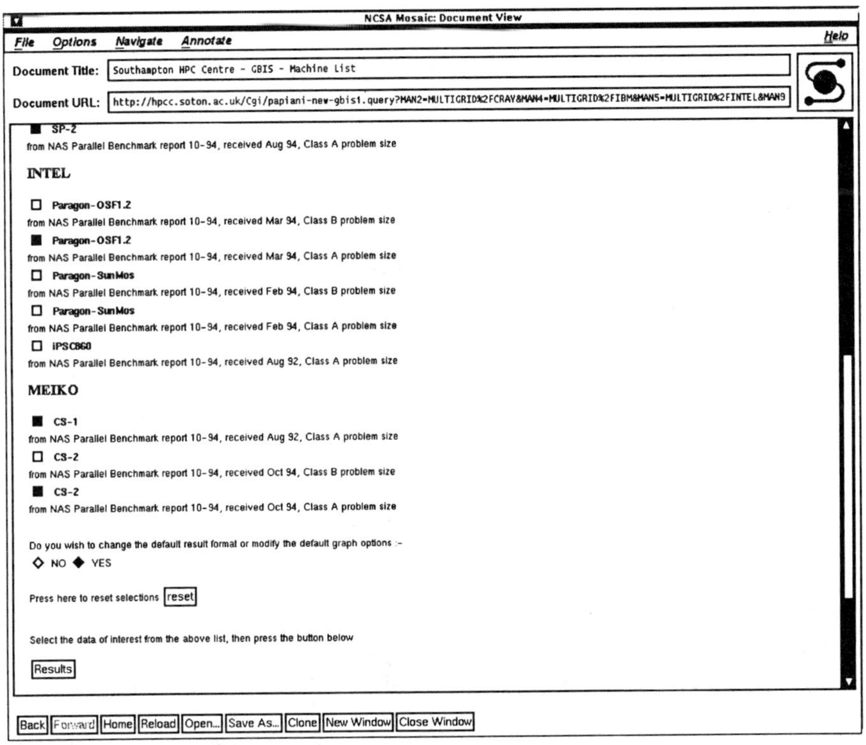

Figure 5.13: GBIS WWW machine list page.

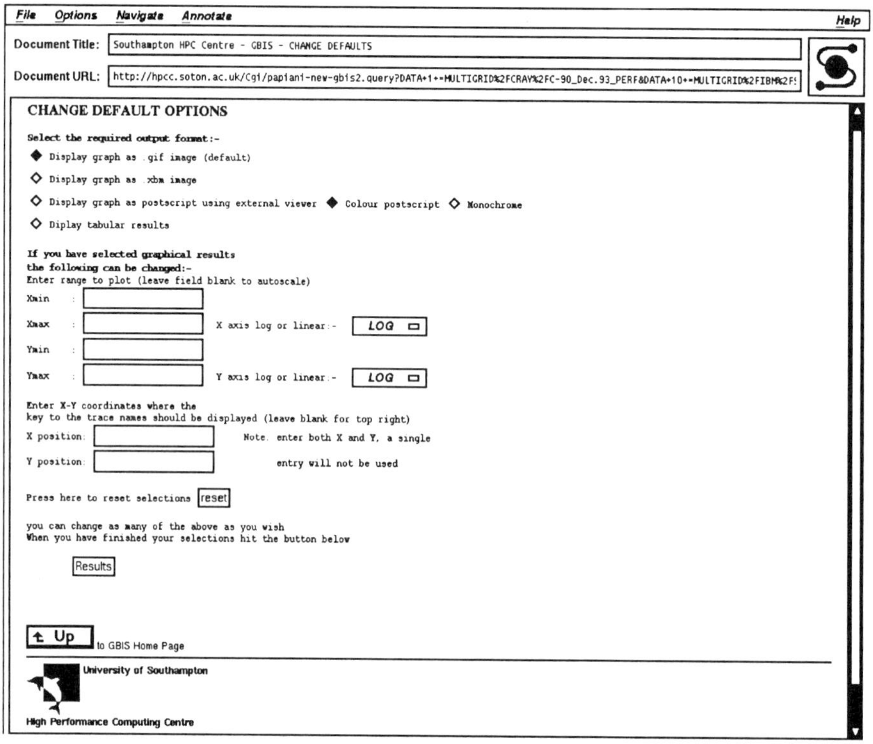

Figure 5.14: GBIS WWW change defaults page.

```
#! CS-2 at the University of Southampton.
## Genesis 3.0, PVM, Fortran 77,
## 26 MAY 94
#
#Message Length / Byte  Transfer Time / us
0                       1.983E-04
1                       2.001E-04
2                       2.006E-04
3                       2.007E-04
#
#RINF=9.180 MByte/s
#NHALF=1791.949 Byte
#STARTUP=195.200us
```

Figure 5.15: Example contents of a GBIS result data file.

As well as lines containing coordinates, there are lines beginning with #!, ## and #. Since each of these lines contains at least one # at the start they are treated as comments by GNUPLOT. The line beginning, #!, is used to produce an HTML file (see section-5.5.4) which is used to construct the Mosaic form which shows the available machine list (Fig. 5.9, step 8). The line contains the computer model name which is displayed next to a selection button on the Mosaic form. Lines beginning, ##, are also included in the HTML file and used to construct the Mosaic form. They are displayed after the machine name to provide additional information to help the user decide whether to select the machine. Lines beginning with, #, are not displayed, but can be used to store other comment information. This information is accessible if text results are requested, since the entire contents of the result data file is displayed in this case.

Bourne shell scripts and PERL programs are being developed to convert automatically the results files generated by the benchmarks into the format required by the GBIS. For example, a PERL program has been written, which takes the Latex version of the NAS Parallel Benchmark Results [10] and creates the required directories and results data files for the GBIS.

5.5.4 Updating the Results Database

Updating the GBIS database involves three steps;

1. Adding the new result data file to the correct benchmark/manufacturer directory.

2. If a new manufacturer subdirectory was created in 1) then running a shell script which updates the list of manufacturers available for each benchmark.

3. Running a shell script to update the list of available machines available for each manufacturer.

The scripts mentioned in 2) and 3) produce the HTML files which are used in the creation of the Mosaic forms containing the manufacturer list for a given benchmark (Fig. 5.9, step 5) and the machine list for each manufacturer (Fig. 5.9, step 7). The script in 2) uses the names of all subdirectories of a given benchmark directory (the subdirectories are named after manufacturers, see section-5.5.2 and hides the HTML file created in the benchmark directory for later use. The script in 3) works through each of the manufacturer subdirectories and hides an HTML file in each one. This file is constructed by extracting the lines, describing the machine benchmarked, from each result data file as described in section-5.5.3.

The HTML files, used in constructing the Mosaic forms, are updated in this manner in order to improve the response time for the GBIS results. Initially, the Mosaic forms listing manufacturers, and then machines for selected manufacturers, were generated entirely at the time they were required. This gave a much slower response time.

5.5.5 Adding State Information to WWW Pages

A problem that had to be overcome, whilst implementing the GBIS, was how to retain information processed from Mosaic forms for later use. An example of the problem can be described by referring to Fig. 5.9.

If the user selects machines from the machine list WWW page of Fig. 5.9, step 8, and accepts default graph options, this is straight-forward to process, because all the information contained in the Mosaic form is used immediately. If, however, machines are selected from the machine list WWW page and the user opts to change default graph options, the following

CHANGE DEFAULT OPTIONS

Select the required output format:-
◇ Display graph as .gif image (default)

◇ Display graph as .xbm image

◇ Display graph as postscript using external viewer ◇ Colour postscript ◇ Monochrome

◇ Display tabular results

Figure 5.16: Part of the GBIS WWW page which allows default graph options to be changed.

problem occurs. The WWW page of Fig. 5.9, step 11, must be displayed, so that the user can change default graph options, but, at the same time, a means of retaining the machines selected in Fig. 5.9, step 8, is required. The machine names can then be processed with the chosen graph options to produce the appropriate output.

The method used by the GBIS to overcome this problem, is to store the information required for later use in temporary files, and then to encode this file name in the HTML used to create a subsequent form. When the information in this form is processed the name of the file is also extracted.

Figure 5.9, step 9, shows that the names of results data files, corresponding to selected machines, are stored in a temporary file. If the user elects to change default options another Mosaic form is displayed in Fig. 5.9, step 11. Part of this Mosaic form, as it is displayed on a WWW page is shown in Fig. 5.16. The HTML used to generate this Mosaic form is shown in Fig. 5.17. Figure 5.16 shows a radio button used to select the required output format. One of the four options; GIF, XBM, postscript or text must be selected.

In Fig. 5.17 the four lines beginning "`<INPUT TYPE="radio" NAME="FORMAT" VALUE='`", are used to create the radio button. To find out what option was selected, the Bourne Shell script which decodes this Mosaic form, checks the value of an environment variable named `$QS_FORMAT` which contains the contents of the `"VALUE="` attribute for the selected radio button. As well as containing the selected output format, this environment variable also has the name of the temporary file encoded in it. Therefore, the environment variable `$QS_FORMAT`, can be tested to see if it contains the string GIF, XBM, PS or TEXT. This establishes the required output format. This string can then be removed from the environment variable, to leave the name of the temporary file. In Fig. 5.17 the text file name has previously stored in an environment variable named, `$PIDFILE`, and so this is appended to the `"VALUE="` attribute.

5.5.6 Availability

The GBIS homepage is available on the WWW, as follows;

`URL: http://hpcc.soton.ac.uk/RandD/gbis/papiani-new-gbis-top.html`

Results data files can also be obtained by anonymous ftp, if the user does not have access to a WWW viewer. This is accomplished as follows, type:

```
ftp par.soton.ac.uk <press enter>
```

at the `name:` prompt type:

```
 <pre>
 <h2>CHANGE DEFAULT OPTIONS</h2>
 <strong>Select the required output format:-</strong>

<INPUT TYPE="radio" NAME="FORMAT" VALUE="GIF$PIDFILE" CHECKED >Display
graph as .gif image (default)

<INPUT TYPE="radio" NAME="FORMAT" VALUE="XBM$PIDFILE" >Display graph as
.xbm image

<INPUT TYPE="radio" NAME="FORMAT" VALUE="PS$PIDFILE" >Display graph as
postscript using external viewer' '<INPUT TYPE="radio" NAME="PSCOLOUR"
VALUE="color" CHECKED >Colour postscript' '<INPUT TYPE="radio"
NAME="PSCOLOUR" VALUE="monochrome">Monochrome

<INPUT TYPE="radio" NAME="FORMAT" VALUE="TEXT$PIDFILE">Display tabular
results
 </pre>
```

Figure 5.17: Part of the HTML document used to generate the GBIS WWW page which allows default graph options to be changed.

```
anonymous <press enter>
```

at the `password:` prompt, type your full e-mail address followed by `<enter>`, then type:

```
cd pub/benchmark_results <press enter>
```

This takes you to the top-level of the results subdirectories. The subdirectories are in the structure described in section-5.5.2.

Results data files can also be sent in to the GBIS by anonymous ftp, type:

```
ftp par.soton.ac.uk <press enter>
```

at the `name:` prompt type:

```
anonymous <press enter>
```

at the password: prompt type your full e-mail address followed by `<enter>`, then type:

```
cd incoming/benchmark_results <press enter>
```

finally type:

```
put result_file_name <press enter>
```

where, `result_file_name` is the filename of your results file.

A readme file in this directory contains further details, including a suggested convention for the filename of your results file. This readme file is also available via the 'Instructions' link on the GBIS home page. It also contains further information on the implementation of the GBIS.

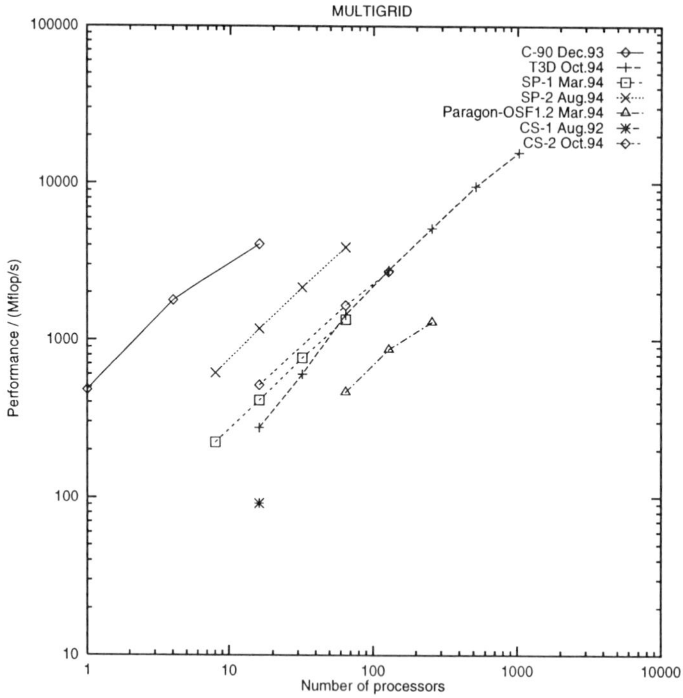

Figure 5.18: GBIS Graph for the Parkbench Multigrid Benchmark (from NASPB), generated using the monochrome postscript output option.

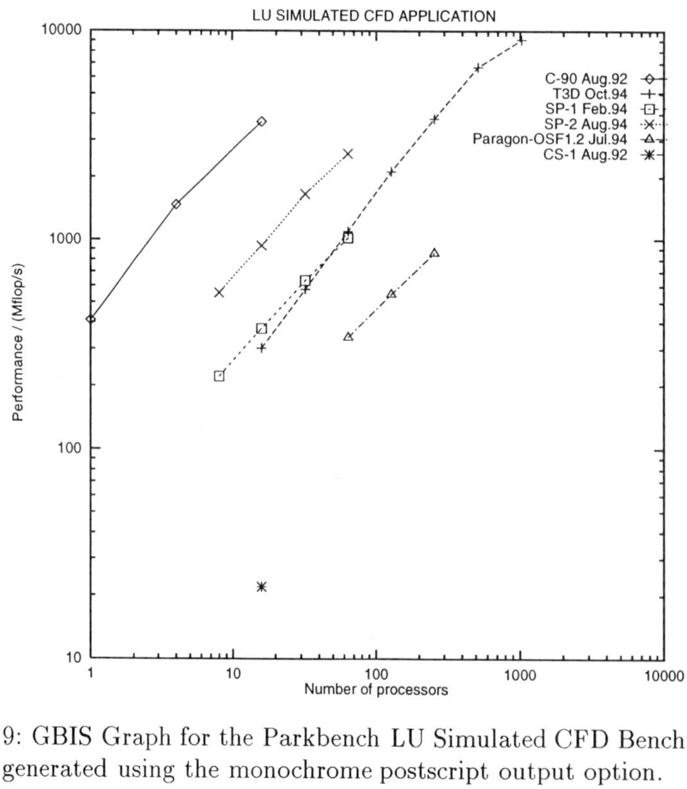

Figure 5.19: GBIS Graph for the Parkbench LU Simulated CFD Benchmark (from NASPB), generated using the monochrome postscript output option.

5.5.7 Example GBIS Graphs

Figures 5.18 and 5.19 show GBIS graphs for two of the NAS parallel benchmarks: Multigrid, and the LU simulated CFD application. This comparison is interesting because, for the first time, an MPP architecture (namely the Cray T3D) exceeds the performance of a top-of-the-range traditional parallel vector computer (namely the Cray C-90). But it takes 256 processors on the T3D to equal or slightly exceed the performance of 16 processors on the C-90 (a ratio of 16 T3D processors to one C-90 processor), and all the warnings associated with the use of large numbers of processors apply (see section-1.3.2). All the other MPPs shown in the figures perform worse than a C-90. Notably the Intel Paragon which even with 256 processors performs slightly worse than an IBM SP2 with only 16 processors and a C-90 with two processors.

The above conclusions are only valid for the data available in 1994 and do not necessarily apply now. They do not, for example, include data for the latest parallel vector computers, such as the Cray T-90, NEC SX4 and Fujitsu VPP500, but they do illustrate the value of graphical comparison using GBIS and the use of absolute performance metrics. Manufacturers are continually improving their computers and software, and the only way that most computer users can keep track of the current situation is by comparing publicly available benchmark results from publicly available databases such as PDS or GBIS, most likely accessed over Internet using a World-Wide Web browsing tool. If this book has a purpose, it is to increase the awareness of these facilities to the computing community, and in particular to encourage computer users to run benchmarks and to contribute their results to one of the databases. Only in this way can the performance of new computers be properly and openly compared.

Bibliography

[1] C. Addison, J. Allwright, N. Binsted, N. Bishop, B. Carpenter, P. Dalloz, D. Gee, V. Getov, A. Hey, R. Hockney, M. Lemke, J. Merlin, M. Pinches, C. Scott, and I. Wolton. The Genesis Distributed-Memory Benchmarks. Part 1: Methodology and general relativity benchmark with results for the SUPRENUM computer. *Concurrency: Practice and Experience*, 5(1):1–22, 1993.

[2] C. A. Addison, V. S. Getov, A. J. G. Hey, R. W. Hockney, and I. C. Wolton. The GENESIS Distributed-Memory Benchmarks. In J. J. Dongarra and W. Gentzsch, editors, *Computer Benchmarks*, volume 8 of *Advances in Parallel Computing*, pages 257–271. Elsevier Science Publ. BV (North-Holland), Amsterdam, The Netherlands, 1993.

[3] G. M. Amdahl. Validity of the single processor approach to achieve large scale computing capabilities. In *Proc. AFIPS 1967 Spring Joint Comput. Conf., vol.* **30**, pages 483–485, 1967.

[4] M. Andreesen and E. Bina. NCSA Mosaic - A Global Hypermedia System. *Internet Research - Electronic Networking Applications and Policy*, 4(1):7–17, 1994.

[5] D. Bailey, E. Barszcz, J. Barton, D. Browning, R. Carter, L. Dagum, R. Fatoohi, P. Frederickson, T. Lasinski, R. Schreiber, H. Simon, V. Venkatakrishnan, and S. Weeratunga. The NAS parallel benchmarks. *Int. J. of Supercomputer Applications*, 5(3):63 – 73, 1991.

[6] D. Bailey, E. Barszcz, J. Barton, D. Browning, R. Carter, L. Dagum, R. Fatoohi, P. Frederickson, T. Lasinski, R. Schreiber, H. Simon, V. Venkatakrishnan, and S. Weeratunga. The NAS parallel benchmarks. Technical Report RNR-94-007, NASA Ames Research Center, Moffett Field, CA 94035, March 1994.

[7] D. Bailey, E. Barszcz, L. Dagum, and H. Simon. NAS Parallel Benchmark Results. In J. Dongarra and W. Gentzsch, editors, *Computer Benchmarks*, pages 225–237. Elsevier Science B.V., North-Holland, Amsterdam, 1993.

[8] D. Bailey, E. Barszcz, L. Dagum, and H. D. Simon. The NAS parallel benchmark results 10-94. Technical Report RNR-94-001, NASA Ames Research Center, Moffett Field, CA 94035, March 1994.

[9] D. Bailey, R. Barszcz, L. Dagum, and H. Simon. NAS Parallel Benchmark Results. Technical Report RNR-93-016, NASA Ames Research Center, Moffett Field, CA 94035, October 1993.

[10] D. Bailey, J. Barton, T. Lasinski, and H. Simon. The NAS parallel benchmarks. Technical Report RNR-91-02, NASA Ames Research Center, Moffett Field, CA 94035, January 1991.

[11] D. Bailey, J. Barton, T. Lasinski, and H. Simon. The NAS parallel benchmarks. Technical Report 103863, NASA Ames Research Center, Moffett Field, CA 94035, July 1993.

[12] D. H. Bailey. NASA Ames' Bailey responds to Kennedy article. Parallel Computing Research, Summer 1995. Published by Center for Research on Parallel Computing, Rice University, Houstson, TX 7005.

[13] G. K. Batchelor. *An Introduction to Fluid Dynamics.* Cambridge University Press, Cambridge, 1967.

[14] R. J. Bell. *SI: The International System of Units.* National Physical Laboratory, HMSO Publications, London, UK, 1993.

[15] T. Berners-Lee, R. Cailliau, J. Groff, and B. Pollermann. World-Wide Web: The Information Universe. *Electronic Networking: Research, Applications and Policy*, 2(1):52–58, 1992.

[16] M. Berry, D. Chen, P. Koss, D. Kuck, S. Lo, Y. Pang, L. Pointer, R. Roloff, A. Sameh, E. Clementi, S. Chin, D. Schneider, G. Fox, P. Messina, D. Walker, C. Hsiung, J. Schwarzmeier, K. Lue, S. Orszag, F. Seidl, O. Johnson, R. Goodrum, and J. Martin. The PERFECT Club Benchmarks: Effective Performance Evaluation of Computers. *Intl. J. Supercomputer Appls.*, 3(3):5–40, 1989.

[17] M. W. Berry, J. J. Dongarra, B. H. Larose, and T. A. Letsche. PDS: A Performance Database Server. *Scientific Programming*, 3:147–156, 1994.

[18] Garrett Birkhoff. *Hydrodynamics.* Princeton University Press, Princeton, New Jersey, 2nd edition, 1960.

[19] P. W. Bridgman. *Dimensional Analysis.* Yale University Press, New Haven, 2nd edition, 1931.

[20] Z. Cvetanovic, E. G. Freedman, and C. Nofsinger. Efficient Decomposition and Performance of Parallel PDE, FFT, Monte-Carlo Simulations, Simplex and Sparse Solvers. In *Proceedings Supercomputing90*, pages 465–474, New York, 1990. IEEE.

[21] K. M. Dixit. The SPEC Benchmarks. *Parallel Computing*, 17:1195–1209, 1991.

[22] K. M. Dixit. The SPEC Benchmarks. In J. Dongarra and W. Gentzsch, editors, *Computer Benchmarks*, pages 149–163. Elsevier Science B.V., North-Holland, Amsterdam, 1993.

[23] J. Dongarra, T. Rowan, and R. Wade. Software Distribution Using XNETLIB Database Server. Computer Science Dept. Technical Report CS-93-191, University of Tennessee, Knoxville, Tennessee, March 1993.

[24] J. J. Dongarra. Performance of various Computers using Standard Linear Equations Software in a Fortran Environment. Computer Science Dept. Technical Report CS-89-85, University of Tennessee, Knoxville, Tennessee, March 1990.

[25] T. H. Dunigan. Performance of the Intel iPSC/860 and Ncube 6400 hypercubes. *Parallel Computing*, 17:1285–1302, 1991.

[26] T. H. Dunigan. Performance of the Intel iPSC/860 and Ncube 6400 hypercubes. In J. Dongarra and W. Gentzsch, editors, *Computer Benchmarks*, pages 273–290. Elsevier Science B.V., North-Holland, Amsterdam, 1993.

[27] J. W. Eastwood, W. Arter, N. J. Brealey, and R. W. Hockney. Body-Fitted Electromagnetic PIC Software for use on Parallel Computers. *Computer Physics Communications*, 87:155–178, 1995.

[28] H. P. Flatt and K. Kennedy. Performance of Parallel Processors. *Parallel Computing*, 12(1):1–20, 1989.

[29] A. Friedli, W. Gentzsch, R. Hockney, and A. van der Steen. A European Supercomputer Benchmark Effort. *Supercomputer 34*, VI(6):14–17, 1989.

[30] V. Getov. 1-Dimensional Parallel FFT Benchmark on SUPRENUM. In D. Etiemble and T. C. Syre, editors, *PARLE'92, Parallel Architectures and Languages Europe*, volume 605 of *Lecture Notes in Computer Science*, pages 163–174. Springer-Verlag, 1992.

[31] V. S. Getov. Performance Characterisation of the Cache Memory Effect. *Supercomputer 63*, XI(5):5–22, 1995. Published by ASFRA, Voorhaven 33, Edam, The Netherlands.

[32] V.S. Getov and R.W. Hockney. Comparative Performance Analysis of Uniformly Distributed Applications. In *Proceedings of Euromicro Workshop on Parallel and Distributed Processing*, pages 259–262. IEEE Computer Society Press, 1993.

[33] V.S. Getov, R.W. Hockney, and A.J.G. Hey. Performance Analysis of Distributed Applications by Suitability Functions. In W. K. Giloi, S. Jahnichen, and B. D. Shriver, editors, *Programming Models for Massively Parallel Computers*, pages 191–197. IEEE Computer Society Press, 1993.

[34] W. D. Gropp and D. E. Keyes. Complexity of parallel implementation of domain decomposition techniques for elliptic partial differential equations. *SIAM J. Sci. Stat. Comput.*, 9:312–326, 1988.

[35] A. Hey, R. Hockney, V. Getov, I. Wolton, J. Merlin, and J. Allwright. The Genesis Distributed-Memory Benchmarks. Part 2: COMMS1, TRANS1, FFT1, and QCD2 Benchmarks on the SUPRENUM and iPSC/860 Computers. *Concurrency: Practice and Experience*, 7(6):543–570, 1995.

[36] A. J. G. Hey. The Genesis Distributed-Memory Benchmarks. *Parallel Computing*, 17:1275–1283, 1991.

[37] High Performance Fortran Forum. High Performance Fortran Language Specification. *Scientific Programming*, 2:1–170, 1993.

[38] W. V. Hobbs. RDB: A relational database management system. Technical report, Rand Corporation, December 1991.

[39] R. Hockney and M. Berry (Eds.). Public International Benchmarks for Parallel Computers, PARKBENCH Committee Report No. 1. *Scientific Programming*, 3(2):101–146, 1994.

[40] R. W. Hockney. Super-Computer Architecture. In C. H. White, editor, *Infotech State of the Art Report, Future Systems 2: Invited Papers*, pages 278–305. Infotech International Ltd., Nicholson House, Maidenhead, England, 1977.

[41] R. W. Hockney. Characterization of Parallel Computers and Algorithms. *Computer Physics Communications*, 26(3&4):285–291, 1982.

[42] R. W. Hockney. Characterizing Computers and Optimizing the FACR(l) Poisson-Solver on Parallel Unicomputers. *IEEE Trans. Computing*, C-32:933–941, 1983.

[43] R. W. Hockney. Parametrization of computer performance. *Parallel Computing*, 5:97–103, 1987.

[44] R. W. Hockney. Characterizing Overheads on VM-EPEX and multiple FPS-164 processors. In G. Paul and G. S. Almasi, editors, *Parallel Systems and Computation*, pages 255–272, Amsterdam, The Netherlands, 1988. Elsevier Science Publ. BV (North-Holland).

[45] R. W. Hockney. Synchronization and Communication Overheads on the LCAP Multiple FPS-164 Computer System. *Parallel Computing*, 9:279–290, 1988.

[46] R. W. Hockney. Performance Parameters and Benchmarking of Supercomputers. *Parallel Computing*, 17:1111–1130, 1991.

[47] R. W. Hockney. A Framework for Benchmark Performance Analysis. *Supercomputer 48*, IX(2):9–22, 1992. Published by ASFRA, Voorhaven 33, Edam, The Netherlands.

[48] R. W. Hockney. A Framework for Benchmark Performance Analysis. In J. J. Dongarra and W. Gentzsch, editors, *Computer Benchmarks*, volume 8 of *Advances in Parallel Computing*, pages 65–76. Elsevier Science Publ. BV (North-Holland), Amsterdam, The Netherlands, 1993.

[49] R. W. Hockney. Performance Parameters and Benchmarking of Supercomputers. In J. J. Dongarra and W. Gentzsch, editors, *Computer Benchmarks*, volume 8 of *Advances in Parallel Computing*, pages 41–63. Elsevier Science Publ. BV (North-Holland), Amsterdam, The Netherlands, 1993.

[50] R. W. Hockney. The Communication Challenge for MPP: Intel Paragon and Meiko CS-2. *Parallel Computing*, 20:389–398, 1994.

[51] R. W. Hockney. Computational Similarity. *Concurrency: Practice and Experience*, 7(2):147–166, 1995.

[52] R. W. Hockney and E. A. Carmona. Comparison of Communication on the Intel iPSC/860 and Touchstone Delta. *Parallel Computing*, 18:1067–1072, 1992.

[53] R. W. Hockney and I. J. Curington. $f_{\frac{1}{2}}$: a Parameter to Characterise Memory and Communication Bottlenecks. *Parallel Computing*, 10:277–286, 1989.

[54] R. W. Hockney and C. R. Jesshope. *Parallel Computers 2: Architecture, Programming and Algorithms*. Adam Hilger/IOP Publishing, Bristol & Philadelphia, second edition, 1988. Distributed in the USA by IOP Publ. Inc., Public Ledger Bldg., Suite 1035, Independence Square, Philadelphia, PA 19106.

[55] Roger W. Hockney and Christopher R. Jesshope. *Parallel Computers: Architecture, Programming and Algorithms*. Adam Hilger, Bristol & Philadelphia, 1981.

[56] B. H. LaRose. The Development and Implementation of a Performance Database Server. Computer Science Dept. Technical Report CS-93-195, University of Tennessee, Knoxville, Tennessee, August 1993.

[57] K. W. Mcenery. The Internet, Worldwide Web and Mosaic - An Overview. *American Journal of Roentgenology*, 164(2):469–473, 1995.

[58] F. H. McMahon. The Livermore Fortran Kernels: A Computer Test of the Numerical Performance Range. Technical Report UCRL-53745, Lawrence Livermore National Laboratory, Livermore, California, December 1986.

[59] F. H. McMahon. The Livermore Fortran Kernels test of the Numerical Performance Range. In J. L. Martin, editor, *Performance Evaluation of Supercomputers*, pages 143–186. Elsevier Science B.V., North-Holland, Amsterdam, 1988.

[60] Robert W. Numrich. Cray-2 memory organization and interprocessor memory contention. In Carl Meyer and R.J. Plemmons, editors, *Linear Algebra, Markov Chains, and Queuing Models*, pages 267–294. Springer-Verlag, 1992.

[61] Robert W. Numrich. Memory contention for shared memory vector multiprocessors. In *Proceedings of Supercomputing '92, Minneapolis, MN, November 16-20*, pages 316–325. Institute of Electrical and Electronics Engineering and IEEE Computer Society, 1992.

[62] Robert W. Numrich, Paul L. Springer, and John C. Peterson. Measurement of communication rates on the Cray T3D interprocessor network. In Wolfgang Gentzsch and Uwe Harms, editors, *High-Performance Computing and Networking, International Conference and Exhibition, Munich, Germany, April 18-20, 1994, Proceedings, Volume 2: Networking and Tools*, pages 150–157. Springer-Verlag, 1994.

[63] O. Reynolds. An experimental investigation of the circumstances that determine whether the motion of water shall be direct or sinuous, and of the law of resistance in parallel channels. *Phil. Trans. Roy. Soc. London*, 174:935, 1883.

[64] M. Papiani, A. J. G. Hey, and R. W. Hockney. The Graphical Benchmark Information Service (GBIS). *Scientific Programming*, 4, 1995. to appear.

[65] *Quantities, Units and Symbols.* The Royal Society, London, 1975.

[66] B. R. Schatz and J. B. Hardin. NCSA Mosaic and the World Wide Web - Global Hypermedia Protocols for the Internet. *Science*, 265(5174):895–901, 1994.

[67] W. Schönauer. *Scientific Computing on Vector Computers.* Special Topics in Supercomputing. Elsevier Science Publ. B. V. (North-Holland), Amsterdam, 1987. series editors: G. Rodrigue, S. Fernbach and G. Michael.

[68] W. Schönauer and H. Häfner. Explaining the Gap between Theoretical Peak Performance and Real Performance for Supercomputer Architectures. *Scientific Programming*, 3(2):157–168, 1994.

[69] W. Schönauer and H. Häfner. A careful interpretation of simple kernel benchmarks yields the essential information about a parallel supercomputer. *Supercomputer 62*, XI(4):63–74, 1995.

[70] H. D. Simon and E. Strohmaier. Amdahl's Law and the statistical content of the NAS parallel benchmarks. *Supercomputer 62*, XI(4):75–88, 1995.

[71] D. F. Snelling. A philosophical perspective on performance measurement. In J. Dongarra and W. Gentzsch, editors, *Computer Benchmarks*, pages 97–103. Elsevier Science B.V., North-Holland, Amsterdam, 1993.

[72] K. Solchenbach. The RAPS Project. In *Proceedings of the Fourth EUROBEN Workshop*, pages 42–50, Budapestlaan 6, 3584 CD Utrecht, Netherlands, January 1994. ACCU.

[73] SPEC Benchmarks Suite Release 1.0. *SPEC Newslett.*, 2(3):3–4, 1990.

[74] Joseph Uniejewski. SPEC Benchmark Suite: Designed for Today's Advanced Systems. SPEC Newsletter, Fall 1989. Volume 1, Issue 1.

[75] A. van der Steen. Practical Aspects of Benchmarking. Notes for Workshop on The Science and Practice of Computer Benchmarking, May 1995. HPCN Europe 1995, Milan, Italy.

[76] R. J. Vetter, C. Spell, and C. Ward. World-Wide Web: The Information Universe. *Computer*, 27(10):49, 1994.

[77] L. Wall and R. Schwartz. *Programming Perl.* O'Reilly and Associates, Inc., Sebastopol, CA, 1991.

[78] R. P. Weicker. A detailed look at some popular benchmarks. *Parallel Computing*, 17:1153–1172, 1991.

[79] R. P. Weicker. A detailed look at some popular benchmarks. In J. Dongarra and W. Gentzsch, editors, *Computer Benchmarks*, pages 107–126. Elsevier Science B.V., North-Holland, Amsterdam, 1993.

Index